D1352933

The Commonwealth Fund Book Program
gratefully acknowledges the assistance
of The Rockefeller University in the
administration of the Program.

The Land That Feeds Us

The Land

John Fraser Hart

That Feeds Us

A volume of
THE COMMONWEALTH FUND BOOK PROGRAM
under the editorship of Lewis Thomas, M.D.

W. W. Norton & Company
New York • London

Copyright © 1991 by
The Commonwealth Fund Book Program
All rights reserved.

Printed in the United States of America.

The text of this book is composed in 10/13 Gael, with the display
set in Hellenic Wide. Composition and manufacturing by the
Haddon Craftsmen, Inc. Book design by Charlotte Staub.

First published as a Norton paperback 1993.

Library of Congress Cataloging in Publication Data

Hart, John Fraser.
The land that feeds us / John Fraser Hart.
p. cm.—(The Commonwealth Fund Book Program)
Includes index.
1. Agriculture—United States—History. I. Title. II.
Series: Commonwealth Fund Book Program (Series)
S441.H2954 1991
630'.973—dc20   90–42720

ISBN 0–393–30950–9

W.W. Norton & Company, Inc.
500 Fifth Avenue, New York, N.Y. 10110
W.W. Norton & Company Ltd
10 Coptic Street, London WCIA IPU

2 3 4 5 6 7 8 9 0

To Meredith

Contents

Foreword

We are in an age when, as never before, urbanization and advances in mechanical and biological technologies—and there are many more to come—have changed dramatically life on the traditional family farms of the United States.

The author of this volume has a comprehensive and sympathetic understanding of those whose livelihood comes from tilling the land and raising livestock on small farms. By talking to the farmers who work the land, he is able to provide us with remarkable insights into the pleasures and hardships that accompany progressive, and often revolutionary, change and the ability of the farmer to adapt.

Dr. John Fraser Hart has undertaken a scholarly geographic investigation of the many forces—social, technological, and political—that have impinged on the life and traditions of those who work on farms in America. His experience, based on personal observations over many years, provides an authentic historical picture of the evolving fortunes of those who, for generations, have spent their lives working on "the land that feeds us."

The advisory committee for the Commonwealth Fund Book Program, which recommended the sponsorship of this volume, consists of the following members: Alexander G. Bearn, M.D.; Lynn Margulis, Ph.D.; Maclyn McCarty, M.D.; Lady Medawar; Berton Roueché; Frederick Seitz, Ph.D.; and Otto Westphal, M.D. The publisher is represented by Edwin Barber, vice-president and director of the trade department at W. W. Norton & Company, Inc. Jan Maier serves as administrative assistant. Margaret E. Mahoney, president of the Commonwealth Fund, who devised the book program, provides constant encouragement and intellectual support to the committee.

Lewis Thomas, M.D., Director
Alexander G. Bearn, M.D., Deputy Director

Preface

Why do you city folks hate us farmers so much?"

Those were the first words U. B. Simoneaux spoke to me when I arrived at his sugar farm near Napoleonville, Louisiana. I told him not to confuse ignorance with dislike—few modern Americans know enough about farming and farmers to be able to hate them. In 1980 only three of every hundred Americans lived on a farm, and only fifteen of every hundred had been raised on one. Taking the kids to visit grandfather's farm used to be one of our great national traditions, but today's grandparents live in a town or city, just like the rest of us.

Few contemporary Americans have ever set foot on a farm, or met and talked to anyone who tries to make a living by farming; and fewer still have actually lived and worked on a farm and understand what makes one tick. Even someone who grew up on a farm may have only a dim, and perhaps quite incorrect, knowledge of modern farming because farming has changed so dramatically over the past three decades.

Farming was the great engine that drove American life from 1620 until 1860, perhaps until 1910. Even as late as 1940, one of every four Americans still lived on a farm. It was a very diverse engine that reflected the environmental diversity of a continent, a diversity that became even greater when different groups of people from different backgrounds with different values identified different opportunities in similar environments and responded to them in different ways.

The basic American farming system was developed in southeastern Pennsylvania, whence it crossed the Appalachian upland barrier to flourish in the Corn Belt. In the dairy areas along the northern fringe of the Corn Belt it was modified in response to environmental constraints. A few well-situated farmers produced milk, fruit, and truck crops for sale to city people, but for most American farmers in the nineteenth century, farming was a way of life rather than a business. They were content to produce enough to feed and clothe their families, with some slight surplus they could sell to buy the goods they could not make or grow on the farm.

The westward march of this farming system finally ground to a halt in Kansas and Nebraska, where the rainfall was too scanty and too unreliable for any crop except wheat. West of Kansas even wheat was too risky, and farming gave way to ranching. Wheat farming and ranching were commercial activities from the very beginning. They specialized in producing large quantities of a few commodities for distant markets.

In the oases of the West a whole new commercial farming system developed that reached its peak in the Central Valley of California. Farmers in the West had to specialize in large-scale production of high-value crops that could stand the cost of transportation to distant eastern markets. Farming in the West is so distinctive that it deserves separate treatment, and I have not discussed it here.

I have included the South, however, because the South has always had its own curious mixture of commercial and way-of-life farming. Many small farms in the South have done little more than support a meager existence for the farm family, but farming has always been a business on the large farms that have specialized in producing cotton, sugar, rice, and other cash crops.

After World War I farming in the eastern United States began to change from a way of life to a business, and this change has accelerated since World War II. The successful modern family farm has had to become a large and efficient business; farmers who used to think in hundreds must now think in thousands. They have had to specialize in doing what they can do best, and they have eliminated their less profitable activities. They have had to learn to manage money and machinery as skillfully as they manage crops and livestock; ulcers have replaced blisters as their principal occupational complaint.

This book explores what farmers do, how and why they do it, and how their activities affect the rural landscape of the eastern United States. The best way to understand farms is to talk to the people who are trying to make a living by farming. I have enjoyed the pleasure of talking to many of them, and I am happy to share my experiences with those who have not been as fortunate as I have been.

A second, and equally important, theme of the book is the geography of farming in the East. Farmers have developed similar farming systems over large areas, because the natural environment over large areas is fairly similar and because most farmers respond in similar fashion to similar environmental opportunities and constraints.

Farmers do not operate in a vacuum. They do things in certain ways because their fathers and grandfathers learned the best way to do them, but they must also be ready to cope with rapid economic and technological change, and they may have to adapt the old ways to new circumstances. The need for change seems constant, but ill-advised change can be disastrous, and farmers are always facing tough decisions.

They know how much farming has changed in recent years, and many feel misunderstood and unappreciated—U. B. Simoneaux is far from alone. The farmers I have talked with were extremely cooperative when I told them what I was up to, because they want their story to be told to an unknowing and possibly uncaring body politic. More than one farmer has told me about the lady on the television talk show who was extremely critical of farmers. The host finally interrupted her to ask where she would get her food if it were not for farmers. "At the supermarket," she said. "That's where I've always gotten it."

My basic field research technique has been the open-ended interview, which has been governed by three fundamental principles. First, I have tried to keep my eyes, ears, and notebook open, and my mouth shut. Second, I try to ask intelligent questions and to listen carefully to what people tell me. Third, I am a gullible skeptic. I believe everything I see and am told, but I test everything I see and hear against all that I know and can find out.

I have supplemented my field observations with historical accounts, contemporary descriptions, bulletins of the U.S. Department of Agriculture and state experiment stations, and any other relevant material I have been able to find in a number of libraries. I have relied especially heavily on the treasure trove of information in the various censuses of agriculture. I have not cited specific works in the text, but I have listed most of my major sources in an appendix.

My route traces the mainstream of American agriculture westward from the seedbed in southeastern Pennsylvania to the western edge of the Corn Belt in southwestern Minnesota, and then it swings back along the milky way of dairy country from Wisconsin to New England, with a glance at the fruit-farming area of southwestern Michigan along the way. From New England the route turns along the Atlantic Coast to the plainsland South, a region with quite a different agricultural tradition, and it visits some of the "islands" of agricultural specialization within the vast sea of dark pine forest that covers most of the region.

John Fraser Hart
August 21, 1990

The Land That Feeds Us

1

The Seedbed

The roots of American agriculture are deep in the soil of Europe, and many of the seeds of American agriculture, both literally and figuratively, were imported from Europe. The fertile limestone plains of southeastern Pennsylvania were the seedbed in which these seeds were planted and nurtured, in which they blossomed, and from which they were transplanted to other parts of the nation. These limestone areas are the only extensive tract of truly good farming land on the eastern seaboard of the United States. To the north the soils of New England are sour and stony, slopes are short but steep, and the summers are uncomfortably cool and damp for grain crops. The climate is warmer in Virginia and the states to the south, but the soils must be heavily fertilized, because they are low in organic matter and have been leached of their plant nutrients.

The settlers who came to the middle colonies found good farming land and a land in which they could feel at home. Summers were hotter and winters colder than in Europe, but both were tolerable. The growing season, between the last killing frost in mid-April and the first killing frost in mid-October, was six months long. Settlers had plenty of time to plant and harvest the familiar grain crops—wheat, oats, rye, and barley—and the new grain crop, corn, which the native Americans showed them how to grow. The land was well watered. Three to four inches of rain fell each month during the growing season, and droughts were rare. Even the plants and animals reminded the new settlers of home. They recognized oak, maple, ash, elm, and beech trees in the hardwood forests that covered the land, and they soon learned the value of hickory, walnut, chestnut, and tulip trees.

The trees gave them logs for buildings, rails for fences, and wood for their fireplaces. Trees were also useful indicators of the quality of the land, because different kinds

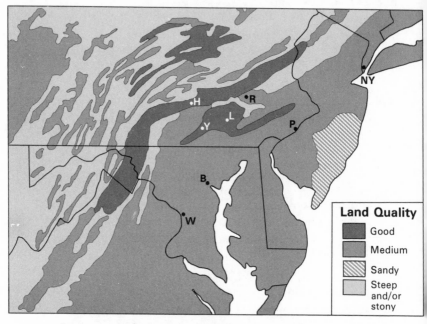

Figure 1–1. The seedbed of American agriculture. The only extensive areas of good farming land on the eastern seaboard of the United States are the limestone plain around Lancaster, Pennsylvania, and the Great Valley northeast and southwest of Harrisburg between the Blue Ridge Mountains and the easternmost ridges of the Appalachian Uplands. The fertile limestone valleys north of Harrisburg were so hard to reach that they were settled relatively late. The letters indicate cities: B=Baltimore, H=Harrisburg, L=Lancaster, NY=New York, P=Philadelphia, R=Reading, W=Washington, and Y=York.

of trees grow in different kinds of soil. The settlers used the trees to identify the best farming areas. They learned to "read" the trees as easily as modern Americans read road signs. The soils near the coast were only middling, but west of Philadelphia was the Lancaster Plain, one of the finest farming areas in North America (Figure 1–1). Local people still call it the Garden Spot. The Lancaster Plain is underlain by massive beds of limestone that weather to form deep, rich, productive soils. The rolling surface is easy to cultivate, and erosion is no problem.

In 1710 a group of German Mennonites began farming in the Conestoga valley near the present city of Lancaster. Wheat was the first commercial crop, as it was in many pioneer areas. Within a generation farmers were hauling wagons loaded with grain to Philadelphia or Baltimore for shipment to Europe, and by 1757 French farmers were complaining that they could not grow wheat cheaply enough to compete with imports from Pennsylvania. Farmers on the Lancaster Plain grew wheat on a third of their cultivated land. They used the rest to grow corn for their cattle and hogs, oats for their horses, and rye for making whiskey. Every farm had a vegetable garden and an orchard of apple, peach, and cherry trees.

A farmer needed three or four horses to plow his land, and a team of four to pull a loaded Conestoga wagon to Philadelphia or Baltimore. Most farmers kept half a dozen motley cattle for milk and meat, fattened a few hogs, and had a flock of barnyard hens for eggs and Sunday dinner. They consumed about half of their produce on the farm and sold the rest either locally or for export.

After the American Revolution farmers on the Lancaster Plain began to shift from wheat to beef cattle. Philadelphia and the other seaboard cities were growing rapidly, and they needed ever larger quantities of meat. Their demands were satisfied, in part, by cattle from the newly settled areas of the west. Professional drovers assembled great herds of cattle in the back country and walked them eastward toward Philadelphia. The cattle lost weight en route, and they were lean, gaunt, and hungry by the time they reached Lancaster. The local farmers bought them cheaply, fattened them on corn, and shipped them off to the slaughterhouses of Philadelphia for a nice profit. The Lancaster Plain became a major beef cattle fattening area, and for many years Lancaster was one of the nation's leading cattle markets.

Farmers on the Lancaster Plain were shifting from wheat farming to cattle fattening at a time when European farmers were caught up in the ferment of the Agricultural Revolution, and many of the new ideas from Europe were filtering across the Atlantic Ocean. In parts of Europe farmers were only reluctantly abandoning an ancient and primitive cropping system. The medieval three-year crop rotation of food, feed, and fallow had remained virtually unchanged since the time of the Roman Empire. In the first year of this rotation farmers grew their principal bread grain, usually wheat, but areas unsuitable for wheat had to rely on rye—an inferior grain. The food grains were sown in the fall, lay dormant in the ground over winter, burgeoned in the spring, and were harvested at the end of summer. After harvest the farmer plowed the ground and prepared it for a short-season crop, barley or oats, that he sowed in the spring and harvested in the fall. Barley and oats are feed grains that are fed to livestock, although the best barley might be malted and made into beer.

Wheat, rye, barley, and oats all are small grains, which the farmer sowed by scattering their seeds broadcast across the land. He did not plant them in rows, so he could not cultivate them with a hoe or plow to get rid of weeds. By the end of the second year a field had usually become badly infested with weeds, and two consecutive crops so depleted the fertility of the soil that in the third year the field had to lie idle, or fallow, cultivated only to kill weeds and other plants that might steal fertility from the soil. After a year of rest it was plowed for a crop of wheat or rye, and the next cycle of the rotation was under way.

The three-field system produced pathetically meager yields. A farmer could expect to harvest only six to ten times the amount of seed he had planted, and he had to save a sixth to a tenth of each year's crop for next year's seed. People expected to starve, perhaps even die, if the crops failed in a bad year, and even in good years there was precious little feed for livestock. The farmer needed an ox to pull his plow and draw his cart, and the family might have had a milk cow or two, but the cattle were pitifully thin and scrawny creatures. Meat was a luxury known only to the rich and powerful. Milk was preserved for winter food in the form of butter or cheese, which had to be heavily salted to keep it from spoiling.

The three-field farming system required few structures. The cultivated fields were not enclosed by fences. Grazing animals were tended by old people or children, who were supposed to keep them out of the fields where crops were growing. Barns were merely large, rectangular boxes with great double doors on both sides. Inside the barn between these doors was a paved threshing floor where workers threshed the grain by pounding it with hand flails. They opened the doors to create the draft that carried away the chaff. The barn was no place for livestock; in fact, it would have been quite unthinkable for farmers to have kept animals in the barn where they threshed and stored grain. They kept their horses in the stable, their cattle in the byre, and their hogs in the sty. They would have been far more likely to have kept their animals in their bedrooms than in the barn.

The Agricultural Revolution changed nearly every aspect of farming. New crops were introduced, and farmers grew them in an improved four-year rotation. They raised bigger and better livestock, developed a new system of mixed farming, and added new structures to the rural landscape.

The new crops were turnips and clover. Turnips are fodder roots that were planted in rows and used for livestock feed. The farmer could plow or hoe the ground between the rows to get rid of weeds, so he no longer had to let the land lie fallow for this purpose. He could graze cattle on the leafy tops of the young plants, or he could dig the mature roots and save them as winter feed for his animals. Clover is a legume that fertilizes the soil by extracting nitrogen from the air and storing it in nodules on its roots. These nodules remain in the soil when the crop is harvested, and they enrich it with nitrogen, one of the essential plant nutrients. A field of clover can be grazed, or the plants can be mowed, dried, and stored as hay. Young clover plants are delicate. They are easily choked out by weeds, and they must be protected by a "nurse crop," usually either barley or oats, until they have established themselves.

These new crops enabled farmers to expand the old three-year rotation into a new and better four-year rotation. They still planted a food grain, the most important crop, in the first year, but they followed it with a root crop such as turnips, or perhaps mangel-wurzels, potatoes, or beets, in the second. In the third year the farmer sowed feed grain and clover seed at the same time. The clover crop would have become established by the time the feed grain was harvested at the end of the year, so in the fourth year the clover could be grazed or cut for hay.

Farmers no longer needed to let a third of their land lie fallow each year both because they could control weeds by cultivating the root crop and because the clover enriched the soil for the vital crop of food grain that followed it. The elimination of the fallow year increased the amount of land under crops by 50 percent, just at the time when the Industrial Revolution was getting under way, when cities were growing rapidly and the demand for food was escalating.

The new rotation greatly increased the amount of winter feed available for livestock. Many farmers traded up from oxen to horses for draft power because one horse can do as much work as three or four oxen, but it requires better feed. More and better feed also enabled farmers to improve the quality of their meat animals by selective breeding. They culled their poorer animals, and bred only those that had desirable traits, such as weight in the edible portions of their bodies.

Farmers valued their animals for their manure almost as much as for their meat because they needed manure to fertilize their soil in the days before chemical fertilizer had been invented. You could tell how good a farmer was by the size of his manure pile. More manure meant richer soils, richer soils meant better crops, better crops meant larger animals, and larger animals meant still more manure. This interdependence of crops and livestock in a tightly integrated farming system is called mixed farming.

The new cattle-feeding system that developed on the Lancaster Plain after the Revolutionary War incorporated many of the innovations of the Agricultural Revolution. Farmers began to grow clover as a soil enrichment crop, and they used it for pasture or cut it for hay. Fodder root crops never became important, however, because they do not like the hot Pennsylvania summers; farmers planted corn, the native American crop, instead. Corn, like fodder root crops, is planted in rows that can be cultivated to control weeds, and it is a far better feed for fattening cattle and hogs than the root crops it supplanted. In fact, corn was so successful that it replaced wheat as the principal crop in a new Pennsylvania four-year rotation of corn, wheat, oats, and clover. Wheat was still a cash crop, and farmers needed oats to feed their work horses, but corn and clover were the feed crops that they used to fatten cattle. These fattened cattle, in turn, produced prodigious quantities of manure that were returned to the soil.

Mixed farming required new structures. The farmer not only had to enclose his fields with fences, hedges, or stone walls to protect his crops against marauding animals and to protect his animals from casual breeding with scrubby neighbors. He also had to shelter the animals against harsh weather. He needed a secure place to store his crops until he was ready to feed them to his animals, and he needed stout pens where he could feed the animals and collect their manure. Finally, he needed a barn where he could thresh and store his grain. Some Swiss farmers satisfied all their requirements under a single roof by building two-story structures, with livestock on the lower level and a threshing floor on the upper level. These two-story "Swiss" barns were models for the new barns that were built in southeastern Pennsylvania.

Around 1780 farmers on the Lancaster Plain began erecting the large, two-level "Swiss" bank barns that are such distinctive features of the rural landscape in south-eastern Pennsylvania. These barns are 80 to 100 feet long, 50 to 60 feet wide, and 50 feet high. A sloping ramp, or "bank," on the exposed north or west side leads to wagon-wide double doors that open onto the upper level. The upper level is a standard European barn, with a central threshing floor and mows on either side. Originally farmers used the mows to store sheaves of grain hauled in from the fields to be threshed, but later they also used them to store hay. Typically the upper level had bins for grain storage and a small tool room and workshop. Straw from the threshing floor was blown out into a huge stack in the stockyard, or strawyard, on the sheltered south or east side of the barn. Cattle were fed and manure was collected in the stockyard, which was enclosed by a stout board fence or stone wall, and sometimes even covered by a roof to protect the precious manure.

One of the most characteristic and distinctive features of the barns of the Lancaster Plain is the forebay, or overshot, which projects out over the stockyard. The entire upper level extends four to six feet beyond the lower level, and overhangs the entrances to the lower level where the workhorses and cattle were housed. These entrances have divided Dutch doors. The top half of a Dutch door can be opened to let in air and light, and the bottom half can be closed to keep in the animals and to keep out the driving rain or snow. Today large cylindrical silos tower beside most barns, but they were not added until a century or so after the barn had been built, often when the farmer switched from beef to dairy cattle.

The idea of putting crops and animals under the same roof in a single general purpose farm building was introduced to North America by German Swiss farmers in southeastern Pennsylvania. It was a remarkable innovation that has become so widely accepted that most Americans take it for granted. They have difficulty realizing that livestock have never been allowed in barns in most of Europe, where the animals are kept in separate structures and the barn is reserved for grain.

The Lancaster Plain was the fertile seedbed from which a crop rotation, a mixed farming system, and distinctive barns were transplanted to southwestern Ohio, where they were slightly modified and became the norm for the agricultural heartland of the continent. The Lancaster Plain is the core of the Pennsylvania Dutch country, which is still renowned for its fine farms and prosperous agriculture.

The Pennsylvania Dutch country might not be nearly so well known today were it not for the conservative Mennonites and their even more conservative cousins, the Amish "plain people," who believe in nonresistance and in separation from the world. Twelve hundred Amish families still farm with horses rather than tractors in the area east of Lancaster. They use oil lamps rather than electric lights, and they send their children to one-room country schoolhouses. They clatter up and down the country roads in horse-drawn buggies.

One of the most basic tenets of the Amish religion is a belief that farming is the only proper way of life. The Amish people believe they can preserve their faith only if they live on the land, near each other and separate from the world. They raise their children to follow in their footsteps and forbid them to live in towns or even to engage in occupations that are not closely related to farming. They teach the children that hard, physical labor should be a pleasure rather than an ordeal, and they work from dawn to dusk. While the Amish are rigorously conservative, they are also good farmers, receptive to new technology so long as it does not go against their religion. For example, they run small engines on gasoline, batteries, or propane gas rather than electricity, which comes in by wire from the world outside. They have shown great ingenuity in adapting modern farm machines to horse drawn operation.

Amish farms are small. A farmer working with horses can handle no more than about 60 acres, which is smaller than many single fields in the Midwest. Nevertheless, land is scarce in the Amish community, because families are large and fathers want to set up each son on his own place. Some groups have reluctantly migrated to other areas and established new colonies, but some Amish farmers on the

Lancaster Plain have tried to solve the problem by dividing their already small farms into even smaller units. Farmers on such small units must specialize in products that can justify the use of large amounts of intensive labor, products such as milk, butter and cheese, poultry and eggs, and tobacco. Even though the Amish raise tobacco, they forbid the use of pipes or cigarettes, but they permit cigar smoking and chewing.

Tobacco became an important crop on the Lancaster Plain after the Civil War for many of the same reasons the Amish raise this cash crop—the scarcity of land. William Penn's original land grants had run to several hundred acres or more, but within a few generations they had been divided among heirs into small farms of only 80 to 120 acres. By 1870 farmers were beginning to feel the pinch of size, and they scouted around for a new money crop. Tobacco filled the bill, and it has remained the leading cash crop in Lancaster County for more than a century. In 1982 tobacco accounted for nearly one-third of the total value of all crops sold in the country, although that figure is somewhat misleading, because most crops in the county are grown for livestock feed rather than for cash sale (Table 1–1).

Tobacco must be dried for several months before it is ready for sale, and the farmers who began growing tobacco had to build special sheds in which they could cure the crop. A typical tobacco shed measures 32 by 54 feet, and 30 feet to the eaves. It is near the farmhouse, and it is painted white like the other farm buildings. Every fourth or fifth board on the sides and ends is hinged four-fifths of the way to the top. The bottoms of these boards can be swung out and propped open for ventilation in good weather, and they can be closed to keep out moisture in foul.

The farmer begins to harvest his tobacco crop around

Labor Day. He chops off the stalk at ground level, upends the plant, and impales the stalk on a wooden lath, or tobacco stick, six plants to a stick. He hauls the loaded sticks to the tobacco shed and hangs them on the lattice work of poles inside it. After the tobacco is cured the farmer must strip the individual leaves from the stalk and tie them into bundles for sale to buyers from tobacco companies, who usually visit the farm in January or February.

Tobacco is a good cash crop, but six acres is about as much as one family can handle, because growing tobacco is backbreaking work. In 1982 a Lancaster County farmer had to work 7 to 10 hours to grow an acre of corn but 150 to 200 hours to grow an acre of tobacco.

Farmers on the Lancaster Plain have intensified their operations to compensate for their small farms after each of our major wars. After the Revolutionary War they shifted from wheat farming to cattle fattening. After the Civil War they began growing tobacco. After World War I they switched from beef cattle to dairy cattle, and they have added poultry and hog enterprises since World War II.

Many farmers on the Lancaster Plain switched from beef to dairy cattle after World War I because they could make more money selling milk to the nearby cities on the East Coast. They fed their cows corn silage and hay, and they needed new structures in which to store these feeds. They cut the corn crop while it was still green, chopped it, and blew it into the new cylindrical silo that towered over the old bank barn. They stored hay in the lofts on the upper level of the barn and enlarged the lower level by replacing the wall on the barnyard side with a new wall that was flush with the upper level. Milking stanchions replaced the old stalls on the enlarged lower level. Since World War II many dairy farmers have shifted to labor-saving milking parlors, with pipeline milking systems and bulk storage tanks. They have invested in airtight, glass-lined, blue metal silos that keep their crops as fresh as they were the day they were harvested. The airtight metal silos also allow them to ensile hay as well as corn.

In 1982 two-fifths of the farmers in Lancaster County sold dairy products, which accounted for a quarter of the county's total farm income (Table 1–1). The average dairy farmer had 40 cows that averaged 13,000 pounds of milk in each 305-day lactation period, but a superior dairy farmer aimed for 18,000 pounds of milk from each of 110 cows. He needed two acres of good cropland to support each cow and the full-time help of a grown son or a hired man. A third of the farmers in the county still fattened beef cattle, but the value of fat cattle sold was only half the value of dairy products sold.

Since World War II many farmers have expanded by adding specialized poultry and hog operations. Sales of poultry and eggs accounted for a third of the county's total farm income in 1982, and sales of hogs added another 10 percent (Table 1–1). The new poultry and hog houses are

Livestock, poultry, and their products	$ 511,651,000
Poultry and poultry products	192,182,000
Dairy products	157,722,000
Cattle and calves	84,805,000
Hogs and pigs	68,835,000
Other	8,107,000
All crops	63,208,000
Tobacco	18,990,000
Corn	13,848,000
Nursery and greenhouse products	13,720,000
Other	16,650,000
Total	$ 574,859,000

Table 1–1   Value of Agricultural Products Sold in Lancaster County, Pennsylvania, 1982

long, low, one-story metal buildings. They are marvels of modern technology, with price tags to match.

All manure is carefully plowed back into the soil. An average acre of cropland in Lancaster County receives about 14 tons of manure each year. The county gets almost $1 billion worth of plant food a year from recycled chicken feed alone. Poultry manure is ideal for corn, but it is so rich that a farmer cannot use more than six or seven tons to the acre. He can use up to 20 tons of cattle manure, if he has it. Such lavish use of manure is creating a water problem. Most farms in the county have their own wells, and more than half are polluted by manure and by silage.

The cropping system in Lancaster County has become more specialized since World War II. Two-thirds of the cropland is in corn, a quarter is in hay, and most of the rest is in wheat or tobacco. There is no longer a standard rotation, but most farmers follow a four-year cycle. In the first two years they harvest corn for grain, and in the third year they cut it for silage. They plant strips of tobacco in the corn fields. After they have cut silage they plant winter wheat, and the next spring they seed alfalfa in the wheat field. Alfalfa has replaced clover as the principal hay crop. The farmer may leave the field in alfalfa for several years if he needs hay, but when the perennial alfalfa starts to decline he plows it up and plants corn again.

Farmers in Lancaster County are scratching for more land. They must expand their operations to maintain their income and to use their equipment most efficiently. The average farmer outside the Amish area owns 120 to 150 acres, and he rents as much more as possible. He would like to buy land, but so would all of his neighbors. In 1982 he figured that a fair and reasonable price for good farmland probably would be in the neighborhood of $3,000 an acre, but at auctions other farmers were bidding the price up to $7,000 or $8,000, and the record price was $10,650 an acre. When you throw in the cost of buildings, equipment, and machinery, you realize that you have to be a millionaire to get started in farming today. The only way a young person can get started is to inherit a farm or to marry one. Many farmers have formed family partnerships or corporations to enable their children to become farmers.

Modern farming has become big business. Farmers have adopted a whole new technology, and they have better crops and better animals and better buildings and bigger debts and more worries and more sleepless nights to show for it. They have to be able to walk into a bank and borrow $1 million or so without batting an eye, and their annual payments of interest alone may run as high as $100,000 or even more.

The contemporary Lancaster Plain is a showplace, aesthetically as well as technologically. Some farms have been in the same family since 1750 or earlier, and some buildings date back to the Revolutionary War. Many farmhouses are massive stone or brick structures with ten to fourteen rooms or more. In front of the house is a neatly trimmed lawn and a luxuriant flower bed ablaze with color, and behind is a garden full of vegetables. Close by is a veritable village of farm buildings: the great bank barn, now much modified, and beside it the old concrete silo and the new blue metal silo; the tobacco shed, its sides scored with ventilator panels; perhaps a new, one-story, metal poultry house or hog house, flanked by a pair of shiny metal feed bins; and a machine shed, a shop, a garage, and other, smaller outbuildings.

The farmsteads are laced together by a maze of narrow paved roads that zigzag through the countryside. There are few fences except for an occasional strand of electrified wire, and at the end of summer the rows of tall corn crowd right up to the edge of the pavement. Blind corners are the norm, and it is woefully easy to get lost if you wander from the main highways. Factories and stores appear at unexpected places along the country roads, and some farmers have sold narrow roadside strips for new nonfarm homes. Conservationists have been concerned about the loss of agricultural land to industrial development and urban sprawl. Perhaps the best way to preserve land for agriculture is to sell it to an Amishman because you know he will never sell it.

The most recent highly lucrative crop that has been cultivated on the Lancaster Plain is tourists—busloads of idly curious city people. The county attracts 5 million visitors a year—300 tour buses and 10,000 passenger cars a day at the height of the season. Traffic jams seem to be permanent summer features on the main roads that run east from Lancaster through the heart of the Amish country, which are uneasy jumbles of eating and lodging places, gift shops and antique stores, factory outlets, museums, and "authentically preserved farms," rides and amusement parks, and luxurious resort complexes.

The first tourists in Lancaster County were attracted by the quaintness of the Amish, but the county has developed into a popular resort area that merely happens to be in Amish country. Some farmers complain that curiosity seekers have become a nuisance, and indeed they have, but the great majority of tourists are afraid to venture far from the main highways. Only a few explore the picturesque old towns and prosperous modern farms in other parts of Lancaster County.

In 1982 I first visited Don Hershey's farm, which is west of Manheim, ten miles northwest of Lancaster, outside Amish country, and well off the beaten tourist track. A sturdy, weatherbeaten sign beside the main road has a picture of a black-and-white cow on a green pasture, and beneath it "Hershvale Farms—Don, Gerry, Larry, Stephen & Patti." A blacktopped lane leads down the slope, past a pond at the lower end of a pasture grazed by Holstein cattle, and up to the white farmhouse on the left. An old 40-by-100-foot bank barn across from the house has been converted into a milking barn with tie stalls for 40 cows, and it has been extended by a one-story 40-by-90-foot cement block addition with 40 more tie stalls. At the far end is a lagoon for manure. Three large cement silos stand by the entrance to the upper level of the barn, which is now used for storing straw and hay and for curing tobacco. Across the lane from the barn is a 28-by-86-foot metal machine shed. Beside it and across from the barn extension is a one-story poultry house that measures 40 by 240 feet.

Don Hershey farms 550 acres of land (Figure 1–2). He

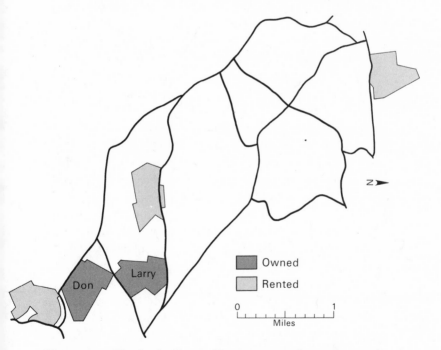

Figure 1–2. Hershvale Farms. Don owns two farms that total 200 acres, and he rents three more that total 350 acres.

milks 110 Holstein dairy cows twice a day, and he also produces and sells eggs, fat hogs, fat cattle, tobacco, wheat, and barley, in addition to milk and purebred registered breeding stock. He was born in 1937, worked on his father's dairy and poultry farm after he graduated from high school, and had various other jobs until 1958, when he rented a 65-acre farm and started farming on his own. Three years later he bought his present 100-acre farm from his father-in-law, and his family still lives in the house where Gerry, his wife, grew up.

"In 1975 I bought a second 100-acre farm," he told me. "It was for sale and I had to buy it, even though I wasn't really ready, because I had to look ahead to the time when my sons would be joining me in the farm business. I also rent three more farms that total 350 acres. I used to rent a couple of other farms but they were sold out from under me. I've been renting additional land ever since I started farming. I find out about land for rent by word of mouth. I pay cash rent for it, and figure that $100 dollars an acre isn't unreasonable."

Don Hershey's cropping program is aimed at producing feed for his animals. Each year he grows 350 acres of corn, and he feeds all of it. He cuts the first 100 acres of corn for silage and expects 18 to 20 tons to the acre. He harvests the rest for grain and averages 100 bushels per acre. He harvests some of the grain when its kernel moisture level is 28 to 30 percent, and he must store it in an airtight silo to keep it from spoiling. He plants 60 acres of barley or wheat undersown with alfalfa on the land from which he has cut silage. He feeds most of the grain, but he sells some for seed, and it brings a premium price. He grows 100 acres of alfalfa and 25 acres of mixed hay, and expects 7 to 8 tons of hay to the acre. He grows 8 acres of tobacco as a cash crop, and the rest of his land is pasture for dry cows and heifers.

Don started off milking 30 cows in 1961 and gradually

built up to 45 in 1977, when he extended the barn. In 1982 he was milking 80. He had to expand his operation because Larry, born in 1959, and Steve, born in 1961, wanted to join him on the farm when they finished high school. "We've formed a legal partnership," he said. "I'm proud that they wanted to farm with me, but we had to have a bigger operation to support three families. I bought a second farm, put up a chicken house, and built a hog house on the second farm. It would have been a whole lot cheaper to send the boys to college. I'm basically a dairyman, but most of my expansion had to be in poultry and hogs, not cows, because there's just not enough land available around here, at least for a reasonable price, to produce the feed we would have needed if we had added more cows. It is also a good idea to diversify your operation, because that helps to average out your income. Dairying, for example, usually gives us a good steady income year after year, although some years it's certainly not the best in the world when you figure in return on labor and investment. The egg business, on the other hand, can be extremely variable, but some years it gives us a return on investment and labor that could never be equaled by milk cows."

Every member of the family helps out with everything, but each one has special responsibilities. Don runs the dairy operation. He and Larry milk 80 cows in the barn at the home place, and Steve milks 30 more at the rented place where he lives. The cows average 15,000 pounds of milk, with a butterfat content of 3.74 percent. They get a standard daily ration of corn silage, alfalfa haylage, and alfalfa hay. Each cow gets a pound of high-moisture corn for each four pounds of milk she produces, and one pound of protein concentrate—roasted soybeans—for each twenty pounds of milk. The milking area in the barn has two overhead vacuum lines. One operates the milking units, and the other sucks milk to the tank where it is kept cool until a tank truck comes to collect it every other day. A milking unit has a head attached to the vacuum lines and four long, slender rubber tubes with tips that clamp on to the cow's teats. Each side of the barn has four units, which is all one person can handle working at full speed.

Don raises his own heifers to replace the old cows that are no longer productive. He fattens the steer calves on corn silage, high-moisture corn, and a protein supplement. He also buys and fattens about 100 lean feeder steers a year.

Steve is the egg man. The poultry house has 31,000 laying hens in wire-mesh cages stacked three high. Each 20-by-24-inch cage holds ten birds. It has automatic feeding and watering systems. "We don't raise any of their feed on the farm," Steve said. "We buy it in bulk and store it in the two large, metal feed bins at the north end of the house. We do our best to keep the birds as happy as possible because happy birds produce more eggs and make more money for us. We control the heating, ventilation, and lighting for the most efficient egg production. Conveyor belts carry the eggs to the packing room. It takes us about six hours a day to pack the eggs. We sell them to a local company. We expect a bird to produce nine eggs every ten days for twelve four-week laying periods, and then we force-moult for four weeks of rest and recovery. We gradually cut back the lights in the house from sixteen hours to eight, which makes the birds stop laying and start shedding their feathers. They rarely get higher than eight eggs every ten days after the forced moult. After the second laying period we sell the spent hens to a soup company and buy a new batch."

Larry is the hog man, in addition to doing much of the field work and helping milk. The hog house, built in 1979, has 17 pens that hold 30 hogs each. At the west end a 20-by-40-foot blue metal silo holds the high-moisture corn on which they are fattened. "We buy lean feeder pigs from a dealer when they are ten weeks old and weigh 40 pounds," Larry said. "We give them a bit of purchased dry feed for the first 30 days, but their basic feed is corn grown on the farm. We fatten them to a market weight of 220 pounds in 120 days. There's a continuous turnover, and we sell around 2,000 fat hogs a year."

The Hershey operation has six tractors, a corn planter, a combine, two manure spreaders, and a full line of equipment for tillage, haymaking, and silo filling. The total investment in land, buildings, and equipment is $1.5 million or more. Don tries to replace two pieces of machinery each year, but he believes that many farmers court financial disaster by buying bigger, better, and more machinery than they need. In 1978 *Pennsylvania Farmer* magazine designated him a Master Farmer, one of only six in a five-state area, but even a Master Farmer is at the mercy of prices, interest rates, the weather, and other forces over which he has no control.

"In 1981," he said, "hogs, steers, and chicken prices all went bad; in 1982 interest went way up; and in 1983 I got hit by drought. It was one of the driest and hottest summers we have ever had in Lancaster County. The southern part of the county had a good thunderstorm on the Fourth of July and got three inches of rain, but I only got a tenth of an inch. In July and August we had 19 straight days without any rain at all. The temperature was up around 100, and out in the middle of the fields it must have been close to 130.

"The corn never even came to tassel; it just curled up. We filled all the silos, but the fields weren't worth putting a sheller into. They had less than ten bushels of grain to the acre. It costs me $50,000 to plant a corn crop, not counting anything for labor, and in 1983 I would have been better off if I had taken the money and gone to Atlantic City and gambled it away. My interest payments were $39,000 in 1981, $85,000 in 1982, and in 1983 I had to borrow $65,000 just to buy feed for my cattle because the drought wiped out my corn crop. You work 14 hours a day, 365 days a year, and these past few years it seems like we have been working for nothing."

Gerry echoed Don's despair. She packs eggs, feeds the calves, takes care of the yard and garden, and, as Don said affectionately, "She's the family go-fer." "I did everything on the farm when the boys were small," she said, "but you look forward to the time when your children are grown and you can start to have the things you couldn't have before and enjoy life. But things are worse now than they ever were." Her voice broke when she said, "I don't even hope any more."

Despite their vicissitudes, however, or perhaps because of them, the Hershey's have deep faith in the Almighty and the same reverence for the land that Lancaster County farmers have shared for two centuries. "All that we have and hope to gain," says Don, "comes only from the merciful hand of God. It is our obligation to take the best care and do the most we can with the talents He has given us. When you're out there working the soil, that's God's soil. When you're out there working with the cattle, they're God's cattle. I may hold title to the farm, but it really belongs to God. I'm just His steward here."

The Great Valley

The outermost ramparts of the Appalachian
Uplands rim the inner edge of the Lancaster Plain. These
uplands are a massive barrier 200 miles wide that extends
southwestward more than 1,000 miles from the Canadian
border to central Alabama. Their slopes are steep, their
soils are thin and stony, and for the most part they are
covered by forests of oaks, maples, tulip poplars, and other
deciduous hardwoods, with spruce and fir at the higher
elevations. They separate the plains along the Atlantic
Coast from the great interior plains that are drained by the
Ohio and Mississippi rivers. For nearly two centuries the
American people remained huddled between the Atlantic
Ocean and the Appalachian Upland barrier. A few brave
souls pushed through to the western waters, but the trend
of the uplands deflected westward movement away from
the west toward the southwest.

The hard hills of Appalachia have few areas of good
farmland (see Figure 1–1). The better farming areas are
underlain by limestone. One of the largest and best is the
Great Valley, which is not so much a valley as a fairly continuous limestone lowland some 10 to 20 miles wide that
curves southwestward from Lake Champlain to Birming-
ham, Alabama. The different sections of the Great Valley
have their own distinctive names. In Pennsylvania, for example, the part east of Harrisburg is called the Lebanon
Valley, and the western part is called the Cumberland Val-

ley. The part in northern Virginia is called the Shenandoah Valley.

The only easy route from eastern Pennsylvania to the Great Valley is the Harrisburg Gap, where the Susquehanna River cuts through the mountains to the Lancaster Plain. Near Harrisburg the Great Valley runs east and west, but 40 miles west of Harrisburg it begins to bend southward into Maryland and Virginia. This bend in the Valley played an important role in the settlement of the United States because it deflected pioneers toward the South (Figure 2–1).

Before the revolutionary war most immigrants landed at Philadelphia. It was the easiest American port for them to reach because the wheat trade required a regular and frequent schedule of sailing to Europe. The main immigrant route to the west led across the Lancaster Plain and through the Harrisburg Gap to the Great Valley (see Figure 1–1), the path of least resistance (Figure 2–1). These pioneers followed the Great Valley westward across Pennsylvania, and then continued "up" the Valley when it turned southward across Maryland into northern Virginia. (In the Valley south is "up" and north is "down;" you "go up the Valley" when you go south, and you "go down" when you go north.)

A few restless and adventurous fellows pushed on westward into the endless mountains of the Ridge and Valley area, and those who found and settled the better limestone valleys were richly rewarded. Others kept searching

Figure 2–1. The bend of the Great Valley. The bend of the Great Valley west of Harrisburg deflected settlers southwestward across Maryland into Virginia. From an original drawing by Erwin Raisz in *The Physiographic Provinces of North America,* by Wallace W. Atwood, copyright 1940 by Ginn and Company. Reproduced by permission of Silver, Burdett & Ginn Inc.

for an easy route through the mountains to the Ohio River and the western waters until they finally discovered Cumberland Gap, where Virginia meets Kentucky and Tennessee.

Two distinct economies and lifestyles developed in the Appalachian Uplands during the nineteenth century. The people who settled the more favored limestone lowlands, such as the Great Valley, became prosperous farmers. These areas were increasingly commercialized and became part of the national economy. The people in the isolated coves and hollows "back up in the hills" made their own livelihood and were largely separate from the rest of the world.

The early settlers in the Valley produced all they needed to feed and clothe themselves, plus a surplus to sell. One of the first surpluses was cattle and horses, which could be droved to distant markets. Cream and eggs were minor sources of income. The farmer kept a few cows for milk, which he let stand until the cream had risen to the top and could be skimmed off. His wife churned the cream into butter, and he mixed the skimmed milk with ground grain to make feed for his hogs. In later years he would sell the cream to a creamery, twice a week putting a large metal can of cream beside the road. A truck from the creamery would pick it up and leave an empty one in its place. The farm wife claimed the eggs laid by the barnyard hens. She carried baskets of eggs to the crossroads store and traded them for "store-boughten" luxuries that could not be produced on the farm.

Wheat was the early settlers' principal commercial crop. By the time of the Revolutionary War farmers in the Valley were already growing more wheat than they could eat, and they sent a steady stream of wagons loaded with wheat, flour, dressed beef, and bacon eastward across the mountains to help feed Washington's troops. Eighty years later the Valley was the breadbasket for the Confederate army. It continued to produce a surplus of wheat until well into the twentieth century, when it could no longer compete with large, mechanized producers in the West.

Farmers in the Valley have grown wheat in regular rotation, before hay and after corn. Wheat has been the money crop, but corn is the crop that has kept man and beast alive. The people ate some of the corn themselves, and they fed the rest to horses, cattle, hogs, and sheep. Corn was essential, but it is hard on the land. Because it is planted in rows and on sloping ground, the bare furrows between the rows are an open invitation to soil erosion every time it rains.

The soil in much of the Valley is thin to begin with, and it is stripped away all too easily by careless cultivation, leaving the gray-white limestone poking up through the topsoil like the bleached bones of some strange monster. Much of the land in the Valley has to be kept in pasture or under hay crops to protect it against erosion. Most of the hay and pasture land is used for beef cattle, which have become the primary source of farm income since World War II. Many farmers also like to keep a few sheep. Sheep do well on the limestone pastures, but they are not as profitable as cattle, because Americans have never developed the taste for lamb and mutton that they have for beef.

The beef cattle business has three distinct stages—breeding, stockering, and feeding—that can be pursued on separate farms. The breeder runs a cow/calf operation: his principal source of income is the sale of calves. He keeps a herd of cows and expects each one to drop a calf each spring. He needs an acre or two of good pasture for each cow.

Calving time in the spring is a ferociously busy season for the breeder because cows can develop all manner of problems when they are trying to calve, especially heifers that are calving for the first time. The rest of the year, however, is fairly easy for the breeder if he has enough pasture. In the summer he has to put up enough hay or other feed for the cows during the winter months, when they are pregnant and must eat enough for two. He can do his other chores whenever it is convenient.

The calves normally are weaned in the fall, when they are six months old, weigh 350 to 400 pounds, and still have a lot of growing to do. The breeder can continue to feed them himself, if he has enough winter feed, or he can sell them as "stockers" to another farmer who has the necessary feed. The stocker farmer will keep the young cattle, now called yearlings, on pasture the following summer and will sell them in the fall at a weight of 700 to 750 pounds.

Lean animals ready to be fattened, or "finished," are called feeder cattle. Feeder cattle are fattened to a weight of 1,000 to 1,100 pounds on concentrated feed, normally corn. Cattle feeding produces much greater returns per acre than breeding or stockering, but it also requires large amounts of concentrated feed. Few farms in the Valley grow enough feed to finish significant numbers of cattle, and most of the feeder cattle raised in the Valley have to be shipped out for fattening.

Breeding and stockering produce returns per acre that are modest at best, and they require fairly large acreages of land to produce a reasonable level of income. Most farms in the Valley are too small to support full-time breeding or stockering operations. They were large enough to maintain a family and to produce a small surplus in the days of horse drawn plows, but they are undersized in the age of tractors.

Few farms in the Valley are large enough to support the people who live on them unless these people have off-farm income or unless they can find some way of enlarging the size of their farm businesses. One way to enlarge the size of a farm business is to engage in a specialized, intensive, form of enterprise. The northern end of the Valley in Virginia has two good examples, apple orchards around Winchester and turkeys near Harrisonburg. Apple orchards and turkeys both require large amounts of labor, and both can provide full-time employment on farms too small for successful beef cattle or field crop operations.

Another way to enlarge the size of a farm business is to acquire more land. Billy Beckner has combined five former farms into a single operation. He was born in 1923 and grew up on a farm ten miles north of Lexington, Virginia. For 20 years he practiced as an ophthalmologist in Hagerstown, Maryland. In 1963 he bought a 220-acre farm, right across the ridge from his father's place, with the idea of building it up and eventually retiring on it. A year or so later an adjacent farm came up for sale, and he could not afford not to buy it because it fit in so well with his original farm. Within three years he had bought three more old, undersized, semisubsistence neighboring farms as they came up for sale, and he had a nice compact unit of 838 acres.

In 1975 the farm required more time than he could give it on weekends, and his work at the clinic involved too much administration and not enough medicine, so he just pulled out and moved to the farm full-time. "I'd rather be here anyway," he told me, when I visited his farm in 1982. "You know what they say: once you get dirt under your fingernails you can never get it out. I like to farm, and I'm a farmer more than a doctor, but I have to work as a doctor to support the farm. I have an office in Lexington, and I keep it open four days a week for five hours a day."

Billy's farm is a commercial beef operation, with a flock of sheep on the side. It nestles in an open limestone valley between two low, shaly ridges. The higher parts of the ridges are wooded (125 acres), and the lower slopes are in pasture (476 acres). About a quarter of the land is cultivable. Billy grows 75 acres of corn, 12 acres of sorghum, 40 acres of alfalfa, 50 acres of orchard-grass hay, and 60 acres of soybeans. He cuts the corn for silage, 25 tons to the acre, and stores it in two 20-by-60-foot silos. In a good year 25 acres of corn will fill one silo. He harvests the rest of his corn for grain, and he expects about 105 bushels per acre. He has recently switched some of his corn land to sorghum, which is cheaper to grow and produces almost as much feed. He also likes to grow a bit of barley, which is better than corn for feeding sheep, but he had no barley in 1982.

Billy is able to establish a good stand of alfalfa by direct seeding, without a nurse crop. He makes three or four cuttings of alfalfa hay each summer, and expects a total of four tons from each acre. A stand normally lasts for five or six years, after which he plows it up, grows corn for a year or two, and then plants alfalfa again, but he has no regular rotation. He is especially proud of his soybeans, which he raises as a cash crop. "They say you can't grow soybeans here," he said, "but I've grown a short-season variety successfully for ten years, and I get 40 to 50 bushels to the acre."

Billy feeds all of his corn and hay to his livestock. He has 125 cows, and he raises enough feed to fatten all of his own calves. Each year he also buys 100 additional heifers when they are a year to eighteen months old and weigh 500 to 600 pounds and fattens them to sell at a 1,000 pounds or so. The cattle graze outdoors from the middle of April until the first of January. During the winter he feeds them hay, and he finishes them on corn silage. He buys and sells all of his cattle through a local dealer.

In 1983 his cattle operation suffered a severe attack of brucellosis. "I bought a heifer that had it," he said. "The guy that sold her didn't tell anybody about it, and I didn't know she had it. In six months she had infected a 66-cow herd, and 32 of them had to go to slaughter. The state pays you for them, but only $50 a head for cows that are worth $500 to $700. You really are just giving them away."

His sheep and cattle complement each other. They will happily graze the same pastures, and the sheep have been his only profitable venture for the past few years, as the price of cattle has not been good. He has a flock of 100 Rambouillet and Suffolk ewes and expects an average of a lamb and a half a year from each ewe because many have twins. The lambs are dropped between November and February, and he fattens them on alfalfa hay and grain. He likes to have them fattened to 110 or 115 pounds in time for the Easter market when the price for fat lambs is best.

Two of the farms he bought had Pennsylvania German forebay barns. "One old barn," he said, "was in such bad shape that every now and then they had to put a chain around it, hitch a tractor to the chain, and pull it back and forth to straighten it up. I've fixed it up and extended the forebay into a good shed where I can feed cattle and sheep." He also uses the other barn for feeding sheep and stores hay in the lofts of both. He has built two new metal sheds, one for machinery and one for livestock. He has three field tractors, a crawler for cleaning manure out of the cattle pens, three forage wagons, a forage chopper with four heads, a combine with grain and corn heads, round and square hay baler, a corn planter, and all the equipment necessary for tillage.

Billy and his wife Wanda have two sons and three daughters. All of them love to visit the farm, but none of them is interested in taking it over from their father. "Last year I lost $30,000 on the farm," he told me, "but this year I hope to break even. I didn't mind losing money that much when the price of land was rising steadily, but now it hurts. In 1963 farmland around here was selling for $100 an acre, but now it's up to $1,040 an acre, because Rockbridge County was written up in some national magazine as a safe place to retire, and since then people have been flocking in to buy land. Prices have just shot up. One of these days I may have to sell the farm because it's costing me too much money, and I have to work too hard at my medical practice to support it. I may have trouble selling the farm as a

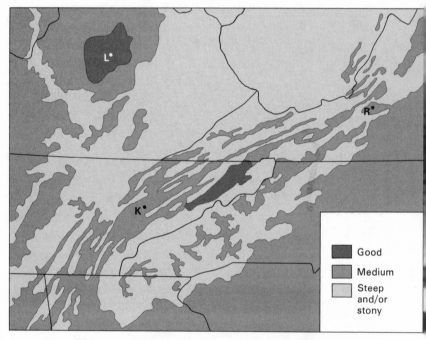

Figure 2–2. Land quality in the southern Appalachian Uplands. Southwest of Roanoke, Virginia, the Great Valley is constricted to a string of nearly separate basins. The only extensive areas of good farming land are in the Watauga country northeast of Knoxville, Tennessee, and in the Bluegrass basin around Lexington, Kentucky. The letters indicate cities: K=Knoxville, L=Lexington, and R=Roanoke.

unit, but I can break it up into the original farms and sell them at a public auction. You couldn't make a living on one of them, but you could use it as a part-time farm."

South of Billy Beckner's farm the Great Valley begins to pinch out, and for 100 miles in southwestern Virginia it is constricted to a series of nearly separate limestone basins, like pearls on a string (Figure 2–2). In Tennessee it widens out again into a belt of good farmland at the foot of the mountains northeast of Knoxville, but the southern end of the Great Valley has thin soils that must be cultivated with great care lest they be washed away by erosion.

The early settlers moving southwestward up the Valley were forced by topography to choose between three routes when they got to Roanoke, Virginia, where the Roanoke River had cut a gap through the Blue Ridge Mountains. Only two major rivers, the Potomac and the James, have breached the Blue Ridge between the Roanoke Gap and the Harrisburg Gap in Pennsylvania, but the mountains are no more than ten miles wide, and they can be crossed by a number of low passes. Southwest of the Roanoke Gap, however, the mountains widen out into the massive barrier of the Great Smoky Mountain system, which is up to eighty miles wide. The flanks of the Smokies are discouragingly steep and rugged, but the high mountain basins in the center have moderately good farmland (See Figure 2–2).

Some of the early settlers moved eastward through the Roanoke Gap and then southward along the North Carolina Piedmont toward Winston-Salem and Charlotte, where their enterprising descendants developed the nation's greatest concentration of cotton textile mills after the Civil War. A second group of early settlers climbed up into the high mountain basins in the Smokies. A third group continued along the line of the Great Valley through southwestern Virginia into northeastern Tennessee, later the staging area for the first great thrust through Cumberland Gap to the vast interior plains of the Middle West.

Not even the people who live there would claim that the hills and mountains of southwestern Virginia are good farming country. Those who settled the area had to abandon the Pennsylvania German farming system. They resorted to the simpler practices that were used on the poorer and steeper flanks of the limestone valleys farther north. Corn replaced wheat as the principal crop and bread grain, and on many farms corn and perhaps tobacco were the only field crop because both corn and tobacco repaid intensive labor on small patches of ground.

In the fall the farmer would cut off the ripe cornstalks near the ground and stack bundles of stalks in shocks scattered through the field. Whenever it was convenient he would break off the ears, shuck them by tearing off the dry outer husks, and store the ears in a simple, rectangular crib of unchinked logs. The crib would protect the ears from predatory animals, and the free passage of air between the logs would dry them. Later, when cribs were made of boards rather than logs, the builder would leave air spaces between the individual boards. The mountain corn crib, with slatted sides, was the ancestor of the corn cribs that became almost universal features on the farms of the Middle West.

The early settlers cleared and cultivated some remarkably steep slopes in an area that abounds in remarkably steep slopes. The rocks sticking up through the turf on many a modern hillside bear mute witness to how hard people had to use the land in order to survive. They scratched the ground for crops until most of the topsoil had washed away, and then they let the hillsides grow up in "pastures" choked with weeds and stippled with brush and saplings. They turned their cattle out to graze on the steeper slopes, and cut hay for winter feed from the gentler ones.

Some farmers merely stacked their hay in the open, but the better farmers built crude, log hay barns that were larger versions of log corn cribs. These farmers might attach a lean-to shelter to one side of their hay barns to shelter their horses, cattle, and equipment. Many log hay barns, especially those on the larger and more prosperous farms, eventually sprouted lean-to shelters on both sides. In time a farmer might add a clutter of other small log outbuildings, such as woodsheds, smokehouses, springhouses, and the like, but he would not need a traditional barn because he had no wheat to thresh. The Pennsylvania German forebay barn virtually disappears south of Lexington, where the Great Valley starts to pinch out and the farming system changes.

The most imposing farm buildings in the hills of southwestern Virginia are the simple rectangular structures of unchinked logs that were built to store hay. Local people call them barns, but to outsiders they look like no more than large sheds because the taper and weight of the logs limit their length to 30 feet or less. A few of these structures had stalls for livestock on the ground floor beneath a hayloft, but more commonly animals and equipment were sheltered in lean-to wings, with the body of the barn filled with hay from floor to rafters.

Farmers pitchforked hay into the barn through a large opening at the gable end until hay forks eased the back-breaking chore of pitching hay. The hay fork was mounted on wheels on a track that ran the full length of the barn's ridgeline and extended out over the spot where the hay wagons were parked to be unloaded. By manipulating a system of ropes and pulleys, the farmer would lower the hay fork onto the wagon, close it around a bundle of hay, lift the bundle to the rooftree, move it along the track, and trip a lever to dump the hay where he wanted it in the loft. The opening in the gable end of the barn had a door that could be closed in bad weather. This opening, and the projection of the hay fork track beyond the end of the barn, were protected by an extension of the barn roof and often by elaborate sides as well.

The principal openings in the hay barns of southwestern Virginia were in the gable ends, rather than on the sides, as in Pennsylvania German forebay barns. Such a change might seem minor, but it reflected a major change in barn function and the creation of a completely new type of barn in North America. This new barn was much better suited to the new farming system, in which corn replaced wheat as the principal crop. The new barns had little in common with the barns of Pennsylvania except their size, and almost nothing in common with barns in Europe, where they were used only for threshing wheat and other small grains. The Pennsylvania German forebay barn is a European threshing barn above a ground floor with stalls for livestock. This barn lost a major part of its function when wheat was no longer the principal crop. The upper level can be used to store hay, but on many farms it is no more than a handy place to keep machinery or just to store junk.

The hay and livestock barns that evolved in southwestern Virginia were ancestors of a type of barn that became common throughout the Corn Belt. These early barns were built in stages: first a central log crib for hay, then a lean-to shed on one side, finally a lean-to shed on the other. The practicality of the final form was widely recognized, and later barns were built from scratch in this style. They were built of boards rather than logs, but they retained the basic form of low sheds for livestock and equipment, flanking a large central hay barn whose principal openings were at its ends rather than on its sides. This type of barn became standardized in the better farming land of the

Great Valley in northeastern Tennessee, an area that became an important source for the settlers who poured
through Cumberland Gap on their way to Kentucky, the
Ohio Valley, and the Middle West.

3

The Hard Hills
of Appalachia

The people who lived in the isolated coves and hollows back up in the hills of the Appalachian Uplands had an existence that was quite separate from the life of the limestone lowlands. These coves and hollows attracted few new people after the initial surge of settlement had swept past, and they had only limited contact with the world outside. For a century or more time quietly passed them by. A simple farming economy gradually replaced the economy of the frontier, but the old ways lingered on, and in many respects the way of life in 1940 was as self-sufficient as the way of life had been in 1800.

Two types of personalities were needed to help tame the American wilderness. First came the frontiersman, adventurous to the point of recklessness, who roamed far and wide with his axe and rifle always at the ready. The frontiersman could not see a tree without wanting to cut it down. He felled the trees, built a cabin of the logs, and planted corn, beans, and squash, all together on a small patch of ground. He kept no animals, but got his meat from the woods. He was more at home in the woods than in the fields, a hunter rather than a farmer. He was a romantic, restless, footloose fellow, full of wanderlust and ever on the go. He would begin to feel crowded when he could see the smoke from another settler's cabin, so he would pack up his meager belongings and strike off into the wilderness once again. Daniel Boone was one of these men. He knew it was time to move on when he saw the first schoolteacher or preacher.

Farmers quickly moved in and replaced the frontiersmen in the better farming areas, such as the Great Valley. The farmers were more stable types, solid, stolid, and cautious. They were quite willing to let someone else do the hard work of clearing the land and then buy him out. They enlarged the cultivated patches of the frontiersmen into fields and grew crops in orderly rotations. Some of their crops they consumed, some they sold, and some they used to fatten animals that they could also consume or sell. When they prospered, they replaced the rude log buildings of Daniel Boone's ilk with better houses and larger barns.

The two types of personality might be found in the same person, and certainly within the same ethnic group, but although stereotypes can be misleading, it is nonetheless convenient and probably not stretching the point too far to associate them with different ethnic groups. Most of the wild frontiersmen were Scotch-Irish and most of the sensible farmers were Germans.

The Scotch-Irish who settled the Appalachian Uplands were descendants of Scottish Presbyterians whom King James had settled in northeastern Ireland to help control the wild and unruly Irish. His Scottish subjects became disenchanted with Ulster, however, and between 1720 and 1776 more than half a million of them migrated to the New World. They landed at Philadelphia and headed westward across the Lancaster Plain, up the Great Valley, and back into the isolated coves and hollows of the Appalachian Uplands.

The Scotch-Irish were a rough-and-ready lot who became the quintessential people of the frontier. They were lean, tough, and as self-reliant as the Indians, whose way of life they quickly adopted. They thought nothing of traveling enormous distances on foot: during the Civil War their grandsons covered ground so rapidly that Stonewall Jackson's troops became known as "foot cavalry." The men disappeared into the woods for months, even years, at a time on long hunting trips. They needed only axe and rifle, powder and lead, knife and tomahawk to live happily in the wilderness.

These people were deeply religious and fiercely independent. They feared God alone, and distrusted all other authority. They cheerfully ignored King George of England, whose royal Proclamation of 1763 ordered them not to settle west of the sources of rivers that flowed into the Atlantic. They further showed their dislike of the British monarch at the battle of King's Mountain, one of the turning points of the Revolutionary War, when they hid behind trees and picked off his redcoats like wild turkeys.

The Scotch-Irish were keen observers and quick learners. They learned about farming and farm buildings from the Pennsylvania Germans, through whose country they passed on their way west. They became familiar with new crops, such as corn, wheat, and clover, and they learned to grow them in regular rotations. They picked up the idea of keeping both crops and animals under the same roof in a single building they had learned to call a barn. They hailed from areas where wood was scarce, where their cabins had been built of stone or turf; they learned from the Germans how to build cabins of logs cleared from the land, and their log cabins became the very symbol of the American frontier.

They learned two more things from Pennsylvania farmers. They learned respect for the high-wheeled Conestoga wagon, which was marvelously well-suited to what passed for roads on the frontier, and they learned that gunsmiths in Lancaster had greatly improved the old muzzle-loading

musket by elongating its barrel, narrowing its bore, refin-
ing its rifling, and adding a grease patch. The standard
weapon of the frontier became known as the Kentucky
long rifle, but most of the long rifles actually were made in
Pennsylvania.

More than from the farmers, the Scotch-Irish learned
from the Indians, who taught them how to scratch out an
existence from the hardwood forest without using plows
or draft animals. The pioneer learned how to clear a patch
of ground for cultivation by "barking," or girdling, the
trees. He would chop off the bark completely around the
base of each tree, killing it and allowing full sunlight to
penetrate to the rich, dark soil of the forest floor.

He would use a hand hoe to mound up the earth into
"hills," and in each hill plant the seeds of the three basic
crops—corn, beans, and cucurbits. The corn stalks would
grow tall, provide food for man and beast, and serve as
supports for a variety of climbing bean plants. The cucur-
bit vines that crept across the ground beneath would produce squash, pumpkins, cucumbers, gourds, and melons.
Later the pioneer would grow a bit of tobacco, perhaps
some potatoes, and some cotton, flax, or hemp for fibers
that his wife could make into homespun clothing.

The cultivated land consisted of small cleared patches
scattered through the forest. Each patch had its quota of
stumps and dead, fire-blackened tree trunks. These
patches were unsightly affairs, but they produced enough
food to sustain life. Corn was the staple food. The pioneers
boiled fresh corn with beans to make a tasty dish of suc-
cotash. They leached dry corn in lye to make hominy, or
they ground it into meal to be baked on the hearth or fried
in a skillet. For sweetening they grew cane sorghum,
squeezed the juice out of the stalks, and boiled it down to
make molasses. They tapped the spring sap of hard maple
trees and boiled it down into maple syrup. They kept an

eye out for "bee trees" from which they could take honey. They collected blackberries, huckleberries, and other wild fruits and berries.

Over time, however, the cultivated patches on the hill farms were enlarged and started to look more like fields, but they remained small. Plows replaced hoes as the principal implements of cultivation. Plows required workstock, either horses or mules, and workstock required feed. Corn was the principal and sometimes only crop. It fed the people and their animals, and some of it was distilled into whiskey. Some farmers also had a tiny patch of tobacco for cash income. Tobacco is a good cash crop for small farms because it requires enormous amounts of labor and produces a large return from a small area. Some farmers cut a bit of weedy hay from the land that was too worn out to grow crops any longer. Nearly every farm had a vegetable garden, a small orchard with a few fruit trees, a milk cow or two, a few hogs, and a flock of chickens.

The pioneer women and children did much of the work of cultivation because the men were off hunting the wild game that put meat on their tables. The pioneer kept half a dozen lean and ugly hound dogs to help him hunt and to warn him against surprise by Indians or other visitors. After game became scarce he turned hogs and cattle loose to forage for themselves in the woods. Hogs did well on the mast—nuts and acorns—they could root out of the forest floor, but they became semi-wild and dangerous, and often they had to be hunted just like other game.

In the fall the pioneer rounded up a few hogs, if he could, put them in a pen, and fed them corn for a couple of weeks to remove the bitter acorn taste from their flesh. At "hog killing time," after the first sharp frost, the hogs were killed, hung by their hind legs to drain the blood from their carcasses, doused with buckets of scalding water and scraped to remove the bristles from their hides, and cut up into hams, shoulders, and the like. Nothing was wasted. The pioneer rendered the trimmings of fat into lard, made the lean trimmings into sausage, and worked salt into the larger cuts to preserve them. He hung these salted cuts in his "smokehouse" and cured them over a smoky fire of green hickory chips. The cure was not completed until spring, and it became traditional to have the first ham of the year on Easter Sunday.

The pioneer and his family grew or made nearly everything they needed. They were as self-sufficient as possible, but it is incorrect to say they had a "subsistence" economy because even from the very first they had to have rifles, powder, axes, and other manufactured goods they could not make for themselves. They satisfied their needs by selling or bartering furs, skins, and cattle in the early days, and later surplus whiskey, tobacco, and other crops.

The old way of life supported a remarkably dense farm population in an extraordinarily difficult physical environment. Through the years the population continued to grow because many sons were needed to work the land. Birthrates were high, and families were large. Few rural areas in the United States have been as congested as the coves and hollows of Appalachia, if one measures population density in terms of the ability of the land to support the people who live on it. Although large numbers have migrated to other parts of the United States, many return when they retire because this is home. Some do not like to leave their homes in the first place, and others remain because of their stern and unyielding religious convictions. They believe that God's will put them where they are and that God will reward them in the afterlife for the privation and suffering they endure on this earth.

The land of the Appalachian Uplands is hard, and it is also astonishingly fragile. Much of it is too steep for successful cultivation, and it should have been left in forest or used only for pasture, but the people had to cultivate it anyway because it was the only land they had. Each community has its joke about the man whose field was so steep that he fell out of it, but when you travel through the hills of Appalachia you begin to wonder whether it really is a joke. The land usually yielded a good crop the first year after it was cleared, but every heavy rainstorm washed away the precious topsoil and far too soon the poor, bare bones of the underlying rock began to stick right up through the eroded hillside. The only land that can be cultivated year after year is the narrow strips of bottomland along the streams, and the bottomlands are inundated by the floods that seem to be a rite of spring.

After a few years of cultivation the hillsides lost their fertility, even if they did not lose their topsoil, and the farmer had to clear a new patch of ground to cultivate. Perhaps he would cut a bit of hay from the old field for a year or two or turn a few cattle loose to graze on the gully-scored pasture, but old Mother Nature quickly foreclosed her mortgage on it, and soon it was overgrown with unpalatable grasses, unclipped weeds, and brush. After the land had rested under "brush fallow" for a while, however, the owner might come back and clear it once again when he ran short of cultivable land. Much of the land of Appalachia has been through the cycle of land rotation (clearance, cultivation, brush fallow, and reclearance) more than once.

The cycle of land rotation (which can just as accurately be called "brush fallowing" or "shifting cultivation") is not nearly so regular as crop rotation, but it involves the same basic principle of allowing the land to rest after a period of demanding use (cultivation) by putting it to a use that is less demanding (pasture or woodland).

The Scotch-Irish settlers in Appalachia were already familiar with the idea of land rotation. In Scotland the only continuously cultivated land was a patch of good soil (the "infield") near the farmstead. The rest of the farm was the rough "outfield" on which the farmer cultivated a new patch of ground each year. He took a crop or two of oats and then allowed the patch to revert to grass until it had recuperated enough for another round of cropping. The land under cultivation in the outfield was rotated, but the crops were not. In the New World oats were replaced as the staple crop by corn, a far better crop, but the cultivated patch was still moved around the farm, just as it had been in Scotland.

One big difference between the Old World and the New was the "bandaids" that Mother Nature used, the plants with which she hid and healed the wounds caused by human use. In the Old World she had relied on grass and a few shrubs, but she had far greater variety in the New. The first plant to colonize fallow land was broomsedge, a coarse, unpalatable, orange-tan grass that was soon followed by blackberry briers, sumac bushes, and tough little pine trees that seemed to be able to take root almost anywhere. Sprouts of sassafras and persimmon began to appear before too long, and in time the field reverted to poor, second-growth hardwood forest. Happily, many of the old field plants were useful. People made brooms from broomsedge, and they boiled sassafras roots to make tea. They picked blackberries in the spring, and in the fall they picked persimmons, a mouth-puckering disaster if you don't know how to eat them.

The cycle of land rotation in Appalachia extends over a generation, or even longer, and rare indeed is the person who knows the entire cycle from personal observation. Few men have had the experience of clearing the same piece of ground more than once in a working lifetime. Even fewer realize that they are reclearing land that had been cleared by their fathers or grandfathers before them.

Land clearance has become much easier in recent years. New machines such as the bulldozer, the brush hog (which can uproot and clear away fairly good-sized brush), and the bog disc (which can "ride" over obstacles and level newly cleared land) have enabled farmers to clear land with far less labor than they needed in the past, and windows of charred, partially burned stumps and brush are a common sight in many parts of Appalachia.

A one-time visitor is far more conscious of land abandonment than of land rotation. Land rotation rarely is mentioned in print, despite its importance as a technique for maintaining production and providing human livelihood in a difficult environment. The length of the cycle means one must visit an area repeatedly and over the years to become aware that forests have been replaced by cornfields, that pines now stand where corn once grew.

At any given time only a small fraction of the land in the hills of Appalachia is being used for crops, probably no more than a quarter is in pasture on its way to brush, and the rest is in brush fallow or second-growth woodland. Most of the land would be in the abandonment phase of the land-rotation cycle whenever a visitor happened to pass through. Anyone who wanted to dismiss the people as benighted and backward could see at a glance that much land had been worn out and abandoned, and these outsiders would be far too quick to jump to the erroneous conclusion that abandonment was forever.

Unfortunately, the ignorance of the rest of the nation has brainwashed the people of Appalachia into feeling that they should be ashamed of a farming system based on land rotation that has failed to produce an economic surplus. No matter that land rotation is an effective technique for maintaining production and sustaining life in a niggardly physical environment. Their values have been ridiculed, and they have learned to avoid scorn by refusing to talk about them.

The people of Appalachia have respected the land as a source of life, and they have not expected it to be a source of wealth. They have used the land for a brief spell, and then allowed it to rest until they have needed to use it once again. They have been content with a small plot of cleared ground and a few livestock. They have been free to come and go as they pleased, to take time off whenever they felt like it, and to feel obligated to no one but themselves and their Maker.

The old way of life has been fading fast since World War II, however, and the amount of farmland in some areas has declined precipitously. Seventeen counties in the heart of the Appalachian Plateau are the extreme example (Figure 3–1). The total area of cleared farmland in these counties dropped from 831,000 acres in 1939 to only 53,000 acres in 1982. Twenty percent of their area was agricultural land in 1939, but only one percent in 1982. The total area of cropland harvested in these counties plummeted from 303,000 acres in 1939 to only 9,000 acres in 1982. In 1982 only two of every thousand acres in these seventeen counties were used to grow crops.

Many small farms at the heads of the hollows have simply been abandoned. These farms were small to begin with, and they became smaller and smaller as they were divided among successive generations. Furthermore, they were even smaller than they seemed because much of the land could not be cultivated continuously, and land that could be cultivated with horses or mules was too steep for tractors and other farm machines.

These small hill farms had to be enlarged in order to provide a reasonable level of income, but farm enlargement is extraordinarily difficult in the hills because most of the good, level land is strung along the creek bottoms in long, narrow strips. Even if the farms could be enlarged, much of the hill land of Appalachia simply would not produce enough to pay for the equipment, fencing, livestock, lime, fertilizer, seed, insurance, taxes, and labor that are needed to farm it. Anyone who has the capital necessary to farm such land would be well advised to buy a good farm somewhere else instead of trying to upgrade a poor one in the hills.

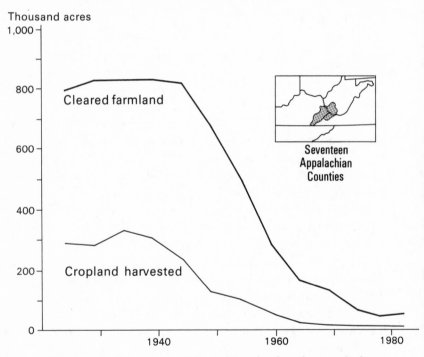

Figure 3–1. Percentage of cleared farmland on the Appalachian Plateau. Twenty percent of a seventeen-county area in the heart of the Appalachian Plateau was cleared farmland before World War II, but a mere one percent in 1982 was. Only two of every thousand acres in these seventeen counties were used to grow crops in 1982.

Accessibility and the age of the owner have been as important as land quality in determining which land would continue to be farmed and which would be abandoned. Some poor land is still farmed if it happens to be near a highway, and better but less accessible land is no longer used if it does not have adequate roads, telephones, electricity, schools, and other appurtenances of modern life. An energetic young man may farm poor land because it is all that he has, whereas an older man who owns better land may have lost interest in farming it.

Much Appalachian farm land has also fallen into disuse because farm income has been pathetically low, and farm levels of living have been depressing. The hills of Appalachia have some of the deepest pockets of poverty in the United States. Many older people have been eking out a meager existence on government payments, such as Social Security and welfare. Their children aspire to higher living standards than their parents ever enjoyed. Military service has exposed many young men to the way of life "outside," and television has kindled the aspirations of people who hitherto had been quite content.

Even the farms that remain occupied often are little more than places of residence for long-distance commuters. New factories have been established in small towns and rural areas, and people commute up to 50 miles or more each way every day to work in them. Several friends will drive to a convenient crossroads, park their cars for the day, and carpool the rest of the way in a single vehicle. The commuter still lives in the old farmhouse, if it is near a highway, or he may have built a new house closer to the road. He may even fiddle around with a bit of part-time farming if he is not too tired after a hard day's work and a long drive home, but he is no longer a real farmer.

4

Wear Cove

Back in 1959 I realized that the old way of life in Appalachia was starting to disappear, and I went in search of small farms where it still hung on. I visited Sevier County, Tennessee, at the foot of the Great Smoky Mountains east of Knoxville, because eastern Tennessee has had more small farms than any other part of the Appalachian Uplands. In 1959 the average farm in Sevier County had 74 acres, of which 26 acres were wooded and 29 acres were in pasture. This average farm had only 7 acres of hay, 4 acres of corn, and six-tenths of an acre of tobacco, the principal source of farm income. Even then more than half of the farmers in the county made more money off the farm than on it, and two of every five worked more than 100 days a year at off-farm jobs.

In Sevierville I talked to Dave Hendrix, the long-time county agent, who reveled in his well-deserved reputation as a colorful character. When I asked Dave about small farms in the county, he snorted and said, "You just can't make a living on a 40-acre farm any more. The tractor has replaced the mule, and it eats cash instead of corn. The family cow and hog have gone out, and so has the orchard and almost the garden, too. The farmer nowadays buys everything from the grocery store, just like you and me."

He told me that many small farms had been able to hang on because they raised tobacco, which provided their major, and often sole, cash income. The government supports the price of tobacco, but it tries to prevent overproduction by limiting the amount that can be produced and sold. Each tobacco farmer is allotted a share of this amount. The individual allotments are so small that they are measured in tenths or even hundredths of an acre.

The allocation of allotments is exceedingly complex. Dave said, "Last year we had to get every durned school-teacher in the county to come in for two or three weeks to help us figure out the allotments." The tobacco allotment is the most valuable part of many farms, and it is guarded zealously. In 1982 allotments were rented for 10 to 25 cents a pound, and one farmer auctioned his off at 90 cents a pound.

Tobacco farming in the hills of Appalachia is unlike tobacco farming elsewhere because each tobacco district produces its own distinctive variety of leaf with its own distinctive production system, labor requirements, and ultimate use. For example, Don Hershey and his neighbors in Lancaster County, Pennsylvania, raise cigar tobacco. The principal type grown in Kentucky and Tennessee is Burley tobacco, which originally was made into chewing tobacco, but now is used mostly in cigarettes. The major tobacco district in the Carolinas and Georgia produces Bright tobacco, a lighter cigarette leaf that is harvested and cured in a different way.

In each tobacco district the labor is backbreaking because much of the work still is done by hand. Tobacco is the last major field crop to be mechanized, and the degree of mechanization varies greatly from district to district. Tobacco has been both the salvation and the curse of the small hill farmer, who has paid the price of an aching back in return for a modest livelihood. No other field crop grown in the United States, taken acre for acre, demands more work than tobacco or makes more money.

In 1959 the average gross return from an acre of tobacco in Sevier County was $875, and the average gross return from an acre of corn was only $42. The principal cost of growing tobacco was the cost of labor, so the tobacco farmer could keep most of the sales price, but other costs ate up most of what he got for his corn. Corn is far less labor intensive. With modern machinery, for example, growing an acre of corn requires less than 10 hours of work, whereas growing an acre of tobacco requires 200 to 400 hours.

Tobacco has the smallest seeds of any commercial crop. A single teaspoon can hold 25,000. They must be planted in specially protected seedbeds, from which seedlings are transplanted to the field. The seedbed should be no more than 6 feet wide, so the farmer can weed it from one side or the other, and it must be 50 feet long for each acre to be planted. It must be protected against wind, weather, and insects. The farmer has to sterilize the soil to destroy any lurking weed seeds or insect eggs. Modern farmers use live steam or chemicals, but old-timers burned off the surface with stacks of wood and old rubber automobile tires, which make an especially hot fire. After the seeds are planted the farmer must put a six- to ten-inch-high frame of wood around the seedbed and stretch a protective cover of unbleached "tobacco cloth," or cheesecloth, across it.

Around the first of June the farmer must pull the seedlings from the seedbed and set them in the field with a mechanical transplanter that requires a driver and two setters. The setters ride on lowslung seats behind each wheel and take turns placing seedlings in the ground. The machine opens a furrow, squirts a shot of water where a seedling is to be placed, and pulls soil around its roots while the setter holds it in position.

The tobacco farmer fights a constant battle against an incredible assortment of insects and diseases that crave the growing plant. The old-timers killed insects by spraying them with arsenic of lead or picking them off by hand. Improved insecticides and fungicides have made the tobacco farmer's lot easier, but he still must maintain constant vigilance, and he must also pray that his plants will not be parched by drought, drowned by heavy rains, scalded by the blazing sun, or stripped of their precious leaves by hailstorms.

In July and August the tobacco farmer faces the thoroughly unpleasant chores of "topping" and "suckering" the crop. The top of a mature tobacco plant is a pink-and-white flower that contains its seeds. The development of seeds diverts the growth energy of the plant away from its leaves. When the flower begins to develop, the farmer must break off the top of the plant in order to get its mind back on its business of producing leaves. Topping a tobacco plant redirects its energy from the seedhead, where it does not pay, back to the leaves, where it does. By topping time the weather has turned hot and muggy, hardly a breeze is stirring, the stalk is tough and hard to break, and the leaves ooze a gummy tar that clings tenaciously to hands and clothing.

The farmer interferes with the sex life of a tobacco plant when he breaks off its seedhead. The plant takes a very dim view of such interference, and it retaliates by producing "suckers." Its stalk has two small axillary buds at the base of each leaf, and these begin to grow into worthless secondary leaves, or suckers, as soon as the top has been broken off. In the old days, the farmer had to go through the field at least once a week to break off the sucker leaves after he had topped his tobacco. Now he can do the job with chemical sprays, which means at least one less backbreaking job for him.

The farmer harvests his tobacco in early September. He grasps the stalk in one hand, bends it to one side, chops it off near the ground with a hatchet, and impales the butt on a tobacco stick that is an inch square and four-and-a-half-feet long. He slips a sharp metal spear point over the end of the stick to make impaling easier. He impales six stalks on each stick, which is then hauled to the barn and hung to cure.

Ripe tobacco leaves have a water content of 80 to 90 percent, and this must be reduced to around 20 percent before they are ready for market. The farmer hangs the loaded sticks in a curing barn that has a lattice of horizontal beams, or tier poles, spaced about four feet apart. At least three men are needed to hang tobacco. The first man passes sticks from the wagon to a man standing on the lower tier of poles, who in turn passes them up to a man who is hanging sticks in the top part of the barn.

Tobacco could be cured with only a roof to protect it from rain if weather conditions were ideal, but they rarely are, so tobacco barns are specially constructed to enable the farmer to control both temperature and humidity. Unlike the tobacco barns on the Lancaster Plain, the side boards of the Appalachian farmer's barn are vertical, and every fourth board is hinged so that it can be opened for ventilation in dry weather and closed during damp spells. Some small farmers cannot even afford special barns, and they hang their crop in any structure that is available.

The farmer must strip the leaves from the worthless stalks after the crop has been cured. He removes the stalks from each stick, tears the leaves from each stalk, and ties them into bundles called "hands." Each hand contains only leaves that have been taken from the same position on the stalk, which determines the quality of the leaf. The farmer sells his tobacco by public auction at the local tobacco warehouse. He is already busy preparing the seedbed for next year's crop before this year's crop is sold.

Armed with knowledge of how to grow tobacco, I went looking for small hill farms in which the old way of life still hung on. Dave Hendrix had urged me to visit Wear Cove, one of the most beautiful places in the eastern United States. Wear Cove is a limestone valley, five miles long and a mile or so wide, an emerald on the northwestern flank of the Great Smoky Mountains in east Tennessee that is completely enclosed by the steep wooded ridges of the Smokies. (Figure 4–1)

In 1959 the old way of life still lingered in Wear Cove. The cove was not completely isolated. You could get in and out if you wanted to, but you had to want to. The

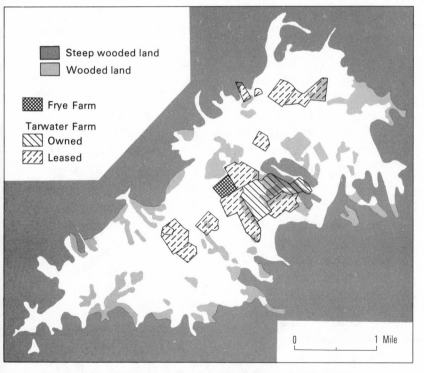

Figure 4–1. Wear Cove. In 1959 Jim Frye made a living on 35 acres in the center of the area. In 1983 C. H. Tarwater, Jr., owned 200 acres and rented another 400 acres from 14 different owners.

narrow road that wound along Cove Creek was not invit-
ing, and few outsiders had any reason to find their way in.
Tuckaleechee Cove, to the south, was more accessible,
and it had already begun to enjoy some of the "blessings"
of tourism. Cades Cove had been sterilized when it was
incorporated into the Great Smoky Mountains National
Park. The people who lived there had been moved out so
the cove could be converted into a picturesque museum
showing the quaint way of life of the people who had lived
there.

Wear Cove remained unspoiled in 1959. Real people still
lived there, and they were living much as their forebears
had lived for a century or more. Their farms were small,
only 40 acres or so. They fed themselves from tiny "truck
patches" of sweet corn, beans, pumpkins, and squash, and
preserved for the winter what they did not eat fresh. They
grew field corn for their animals and tobacco as a cash
crop, but the thin limestone soil had to be used with care,
and much of the land was in pasture for dairy cows and
beef cattle. Most farmers had a small shed for work mules
and hay storage, a smokehouse where they cured the meat
they had butchered, and a springhouse that sheltered the
spring where they got their drinking water. Crocks placed
in the cool water were the "refrigerators" where fresh
food was kept.

Jim Frye's farm was in the heart of Wear Cove. A fresh
bearskin was drying on the wall of his shed. His eyes
squinted suspiciously out at the world from beneath the
brim of the battered brown fedora that was pulled low
over his deeply lined and weatherbeaten face. Eventually
he decided that I was only a harmless professor, and it was
all right to answer my questions.

When I spotted the fresh bearskin I had jumped to the
romantic conclusion that Jim hunted bears to supplement
his food supply, but he snorted that he would just as leave

eat shoe leather as bear meat. Later one of his neighbors told me that Jim and his sons were great bear hunters. They had a camp high on the ridge back of the farm, and it was said that they had shot more than 300 bears.

"Them bears eat hogs sometimes," he said, "and they get into the cornfield or the apple orchard, but otherwise they don't bother us none. We hunt 'em for fun, jes like we hunt squirrels and rabbits and coons. They's thirteen [dead] bears come offa that mountain last year. The boys round here got 'em a gang o' bear dogs, twelve of 'em. Them bears are jes as smart as you are. As soon as they hear them dogs they head for that there [Great Smoky Mountains National] Park, 'cause they know jes as well as you do that you can't hunt 'em in there. Hit takes a good dog jes to keep up with a bear on level ground. Them dogs'll run a bear ten, twelve miles across the mountains, through the roughest places it can find. In summer a man can get et up in them mountains by snakes, but as long as you carry a light at night a snake won't bite you.

"Right now I get more time'n most anything else. I'm not as gainfully occupated some times as others. I have lived in this Valley all my life—65 years if the Lord lets me live 'til September. I been out six or eight months or a year at a time. I go to Sevierville once or twice a week, and jes about every Saturday. My wife is from Sullivan County. I had twelve children, but lost one in infancy, and one was run over by a truck, right up there at the crossroads, when he was sixteen. The others are scattered from here to yonder. I worked 30 years as a tenant farmer afore I bought this place. I got around 35 acres. They's about 8 acres of woodland, and I been a-pasturing about 5 acres. I got some land lays up along this creek here I never tried to get at with a plow, cause this creek gets all over sometimes in a great big s-s-s-floosh. And over there hit's jes a big section of rock. Hit jes comes up, ain't connected up nowhere fer's

I can see. I got about three and a half acres of alfalfy, and about seventeen acres of corn, the biggest corn crop I ever did get out. They's about seven-tenths of an acre of tobacco. Oh, an' about an acre and a half of truck patches and gardens.

"Them truck crops I mostly raise to eat—sweet corn, beans, squash, pumpkins, potatoes, tomatoes. Corn, we feed some of it, and I've sold some. Most years I don't grow too much. Hit jes turned in this year I had this ground I figured on sowin' in pasture, and I planted corn on hit first. We plow corn three–four times, and pick along about the first of November. In the old days we jes pulled hit by hand and throwed hit in a wagon, but now I hire a cornpicker. That sure saves a lot of sore hands.

"Forty years ago folks round here jes raised corn and wheat. Ragweeds would grow up in the wheat stubble, and they'd make hay out of it. They made about 15 to 20 bushels of corn and 6 to 8 bushels of wheat. Now I make about 50 bushels of corn. I cut alfalfy and put hit up for feed. The first cut is in the middle of May, another in 25 to 30 days, and the third in the middle of July. I had most of my hay baled the last few years.

"I got a team of mares I do most of my farmin' with. I got one cow givin' milk and a cow and a heifer comin' fresh. We keep the milk cool in that springhouse there. We used to have five cows, but my wife wasn't able to tend to 'em too much, so I got rid of two of 'em. I keep four head of hogs jes fer meat. I bought 'em, one here in the Valley and three in Knoxville. I gen'ly kill 'em about Thanksgiving and cure the meat in that smokehouse there, over the springhouse. I got about 75 chickens fer meat and eggs.

"I got a disc harrow, a turnin' plow, and a two-horse riding corn cultivator. I gen'ly get somebody, usually C. H. Tarwater, my neighbor 'cross the creek, to mow my pastures with his tractor. Plowin' they charge you $2.50 to $3 an hour; $5 an acre to pick corn, $1.50 an acre to mow, $15 to transplant six-tenths of an acre of tobacco.

"Tobacco's my cash money crop. I rent fourteen-tenths of an acre of tobacco on three different places, six-tenths on one place, six-tenths on another, and two-tenths on the other. I've growed as much as three acres, and I've had as much as two acres right here on this place. I've growed tobacco on a patch of ground has been in tobacco continuin' round twenty years. You can rent land fer, well, the man will pay half the fertilizer and give you half the tobacco. He furnishes the land and you furnish the labor.

"In my case I've been renting for ten years from a man that preaches on Sunday and teaches school all week. He don't bother me none, cause he don't know nothin' 'bout tobacco, and I honestly don't think he likes to work. I furnish everything but the land and the barnyard manure, but it's a piece of land I can get to right next to my own. He never comes about tellin' me nothin'.

"In January, if they's wood a-fit, hit's a good time for burnin' off ground for plantin' a tobacco seedbed around the first of February. You set out tobacco along from the last of May til the tenth of June. We used to set out by hand, but now we hire a transplanter. Hit's a two-row setter. They's about a dozen of 'em in the valley. When you get a stand—and you may have to replace some—you cultivate it three–four times, jes like corn. You got to watch about the worms—spray it with arsenic of lead or pick 'em off by hand. You got to keep the leaves pizened. We top along about 20 July, and sucker all during August. Lots o' people don't like to top and sucker tobacco cause hit's hard dirty work.

"We cut it last of August. Hit takes a tremenjus lot o' barn room. My tobacco barn is 40 by 85 feet. I rent some land from a man that don't have no barn room. We commence gradin' it sometime in October. When it comes 'in case' in damp weather we take the leaves off the stalks and tie them into hands. In January, when I got something like 1,000 pounds worked off, I take hit in and sell hit in Knoxville if the market is strong. Fore you got one crop graded off and sold you're commencing to put out another. Tobacco, hit's a thirteen-month-a-year job. Most o' the folks farmin' in the valley are old men like me. The young uns are leavin'. They want more'n one payday a year, and that not very reg'lar."

In the spring of 1983 I went back to Wear Cove to find out what had changed since 1959. The road that snaked along Cove Creek had been upgraded into a good, modern, highway. New houses dotted the Cove, which had a dense rural population. Few of the residents are "outsiders." Nearly all were born in the Cove. They commute to jobs outside, but they want to continue to live at home on the land they have inherited, which has been divided into ever smaller parcels. Jim Frye's farm, undersized though it was, had been divided among three of his daughters, who rent their land to neighbors. It would have been nice if they had rented to C. H. Tarwater, the neighbor across the creek from whom Jim had hired machines, but the world is rarely so neat and tidy.

Mr. Tarwater was generally recognized as one of the best farmers in Sevier County, and his son, C. H., Jr., 57, has followed in his father's footsteps. "My full name is Clarence Harrison Tarwater, Jr.," he told me, "but nobody has ever called me anything but 'C. H.'" Like his forebears, he knows that he must keep expanding the size of his farm business in order to maintain a reasonable level of income. "My great-grandfather settled here around 1850 on 35 acres. My grandfather built it up to 70 acres, and my father got it up to 300, but I had to sell half of it to pay the estate taxes when he died. I got it built back up to 300, but then I had to sell 100 acres to the government for the new Foothills Parkway they plan to build through here. They lease it back to me, and for all practical purposes it's mine until they start work on the road. I also lease another 300 acres, so I'm farming a total of 600 acres here in the valley."

C. H. experimented with various intensive enterprises before he settled on the egg business as his major farm enterprise. He started selling eggs to the resort hotels in Gatlinburg, which is just across the roadless ridge, but 16 miles away unless you happen to be a crow. "At first we sold mainly to two big hotels, but now we have a regular route and sell eggs to 160 stores and restaurants all the way from Gatlinburg to Sevierville. In 1953 we sold eggs for 65¢ a dozen, and last year they were 64¢. Our labor costs in 1953 were $1,200, and last year they were $85,000.

"I used to have 50,000 laying hens, but I'm down to only 10,000 now. My houses, which were built in 1953, are no longer good enough. I buy eggs from two other farmers, one 30 miles north of here and the other over in middle Tennessee. I used to furnish the hens and feed and pay them six cents a dozen for eggs, but now I buy them at a percentage of the going market price. This can be a pretty cutthroat business. One of the big egg companies came in here and offered eggs to some of my best customers at the market price less 25¢ a dozen, but so far my customers have stayed with me."

In 1982 C. H. raised 125 acres of corn, which yielded 135 bushels an acre. He feeds all of his corn to this hens—he has his own hammer mill to convert grain into chicken feed—but has to buy additional feed. His feed costs were $1,000 a day when he had 50,000 laying hens. He normally grows 15 acres of tobacco. In 1982 his crop yielded 3,600 pounds an acre, nearly double the county average. The rest of his farm is under grass and trees. The pastures merge unfenced into the woods on the steep stony land at the back of the farm. C. H. has underseeded some of the woods to improve the grazing, and he has planted pine trees in a few places, but his cattle graze mainly in the open pastures.

He used to keep a herd of cows and sold their calves, but disease problems became so bad that he switched to steers. He buys 250 to 300 steers in early September, runs them on his pastures for a year, and sells them the following August. The pastures get only manure from the poultry houses, no commercial fertilizer, but C. H. is proud of the fact that he carried 140 steers on 70 acres of pasture last summer. "I cut hay from about 200 acres of pasture in the spring when the cattle can't keep up with the growth of new grass, and feed it to them during the winter. I stack hay in the field. It doesn't look too pretty, but I can do it all myself sitting on that tractor, and it saves labor.

"In recent years I've gone more to pasture and away from row crops, partly to get more of my land grassed down to protect the thin topsoil, and partly to cut back on labor. I used to have eight workers, but I've cut back to three full-time and one part-time, and I'm trying to scale back to what my son and I can handle all by ourselves. My son runs the egg route."

C. H. leases 400 acres in 14 tracts scattered up and down the Cove (Figure 4–1). "I lease these small farms mostly as a favor to relatives or friends," he said. "I don't really need the land, and I'm not making any money on it, but the owners need the rent money, and I'm happy to try to help them." C. H. is keenly aware of his responsibilities as a community leader, and so many of his neighbors depend on him that he is far from free to do as he chooses.

He told me that it would be foolish to invest in any new farm buildings because the threat of development from Gatlinburg hangs like a pall over the entire Cove, and he knows that he probably cannot continue to farm here. Gatlinburg is the entrance to the Great Smoky Mountains

National Park, which has three times as many visitors as
any other national park. The town is crammed into a trian-
gular wedge of relatively level land along the Little Pigeon River. The level land has been completely built up,
and developers have even attacked some incredibly steep
slopes. A whole new ski resort complex, complete with
chalet village and Astroturf for summer skiing, has been
developed in the hills to the west.

Gatlinburg is hemmed in by the park boundary on the
south and by steep mountains on the other three sides, so
the closest level area suitable for commercial development was the bottomland along the Little Pigeon River,
six miles to the north. In 1958 this bottomland still grew
corn. By 1983 it had erupted into the kind of place that
gives tourism a bad name: a continuous roadside strip, 4.6
miles long, of motels, eating places, souvenir shops, and
commercial entertainments such as water slides, rides,
waxworks, "museums," horror shows (intentional and otherwise), and all manner of other grotesqueries. To illus-
trate the importance of tourism in the area, I compared
the numbers of certain types of businesses listed in the
yellow pages for Sevier County (mainly Gatlinburg and
Pigeon Forge) with the numbers listed in two Kentucky
Bluegrass counties (Bourbon and Scott) that have almost
the same area and population (Table 4–1).

Pressure on Wear Cove is going to become far more
intense when construction gets under way on the Foothills
Parkway, a road that runs along the western edge of the
Smoky Mountains. Most of the people in the valley want to
keep their home the way it is. They do not want to sell
their land, but they may not be able to resist the blandish-
ments of developers, because the contrast between farm
prices and development prices is staggering.

C. H. Tarwater told me that a reasonable rent for farmland in the valley would be $20 to $25 an acre, plus 10¢ to

25¢ a pound for tobacco allotment. The traditional rule of thumb says that the price of farmland should be twenty years' rent, so farmland in the valley should sell for $400 to $500 an acre, plus tobacco. The 200 acres of the Tarwater farm are probably worth $80,000 to $100,000. The tobacco allotment of 54,000 pounds would add $5,400 to $13,500 to the rental value or $108,000 to $270,000 to the sales value of his farm. A reasonable estimate of the agricultural value of the farm, therefore, would be in the range of $200,000 to $400,000.

A price of $2,000,000 to $2,500,000 for the Tarwater farm would not seem at all out of line to a big developer, however. People have already been offering $10,000 an acre and up for land in the valley. A solid block of 200 acres would be exceptionally attractive to a big developer, who would be spared the nuisance of trying to buy many small bits and pieces, with the reasonable assurance that the final holdout owners would have figured out what he was up to and would demand a truly outrageous price.

	Sevier	Bourbon and Scott counties
Motels	213	12
Campgrounds & RV parks	32	2
Restaurants	148	43
Gift shops	139	13
Amusement places	49	0
Art galleries & dealers	33	0
Realty companies	32	24
Service stations	59	45
Totals	1,302	723
Population, 1980	41,418	41,218
Area (square miles)	597	584

Table 4–1 Number of Business Establishments in Sevier County, Tennessee, and in Bourbon and Scott Counties, Kentucky

C. H. figures that "Wear's Valley is just settin' for a burst. The developers have been on top of my neck real strong," he said, and he knows that one of these days a developer is going to come along and offer him $2.5 million dollars or more for his farm, ten times what the land is worth for farming. How can he possibly say no? But how can the squire of the valley possibly sell a farm that has been in his family for four generations?

As I was getting ready to leave his place, C. H. looked at me very seriously and said, "Before you go, I want to ask just one favor of you." I wondered what it was going to be, but I assured him that I would be delighted to do whatever I could. "Pray for me!" he said. "I know I can't stay here, but I don't want to leave. This is my life, and I want to stay here, but I doubt that I can. Maybe I could find some land somewhere and run cattle on it and try to build it up, but do I want to try to start fresh in another area at the age of 57? My family has lived here for four generations, and I don't want to see the valley go the way of Pigeon Forge. A lot of people in the valley are watching me to see what I do, and the whole valley is going to go if I sell out, but how can I possibly stay here? I can't sleep nights for worrying about what I should do.

"Please pray for me!"

5

Limestone and
Livestock

Wear Cove is one of the smallest of the open limestone lowlands that are scattered through the Appalachian Uplands in sharp contrast with the wooded hills around them. Many other limestone areas, such as the Nittany and Kishacoquillas valleys in central Pennsylvania, the Big Levels near White Sulphur Springs, West Virginia, Burkes Garden south of Bluefield, Virginia, and the Sequatchie Valley west of Chattanooga, Tennessee, are locally renowned, as well they should be, for the gentleness of their topography, for the goodness of their soil, and for the productiveness and prosperity of their farms, although, like Wear Cove, they may be too small to be shown on generalized maps.

Limestone plains comprise a much larger proportion of the Interior Low Plateau of western Kentucky, middle Tennessee, and northern Alabama (Figure 5–1). The Interior Low Plateau is actually a complex of level limestone basins and plains separated by upland areas that range from rolling to hilly. The upland areas are underlain by a variety of rock types. The upland soils are generally inferior: they are stony, and they are too shallow for sustained cultivation.

The principal areas of good limestone soil are the Bluegrass Basin of north-central Kentucky, the Pennyroyal Plain along the Kentucky-Tennessee line, the Nashville Basin of middle Tennessee, and the limestone valleys of northern Alabama (Figure 5–1). The farms in these areas are larger and more prosperous than those in the neighboring hills, but even in these favored areas the soils must be used with great care because they are thin and easily eroded. Much of the land is kept in pastures that are grazed mainly by beef cattle. The beef farms are mostly breeder or stocker operations because few farmers can raise enough feed to fatten their own animals.

Crops are more important than livestock in the limestone valleys of northern Alabama, one of the last tattered remnants of the old Cotton Belt. In 1982 the area between Florence and Huntsville had a quarter of the entire cotton acreage of the United States east of the state of Mississippi. The other major crop in the area was soybeans, and together cotton and soybeans accounted for more than 80 percent of all its cropland. Northern Alabama is also an intensive poultry-producing area. It relies heavily on cheap feed grain brought from the Corn Belt by barges using the nine-foot navigation channel that the Tennessee Valley Authority (TVA) maintains on the Tennessee River.

The Nashville Basin has thinner soils than the other three limestone lowlands. Much of the land cannot be cultivated, and it must be kept under pasture for cattle, mostly beef but some dairy. One of the most distinctive and characteristic features of the Nashville Basin is its numerous cedar glades, open to dense stands of red cedar on areas of nearly bare limestone. On good soil the cedars would be shaded out by oak, maple, hickory, and other hardwood trees that cannot grow on the bare rock of the glades.

The Pennyroyal Plain of southwestern Kentucky has the most diversified agriculture of any of the limestone areas. Beef cattle are the leading type of livestock, but many farmers grow corn and feed part of their crop to hogs. Soybeans and corn occupy the largest acreage, but tobacco is an important cash crop.

Most farmers on the Pennyroyal Plain raise the same kind of Burley tobacco that is grown in eastern Tennessee and in the Bluegrass, but the western section is a distinctive tobacco district known as the "black patch." The black patch produces a dark, heavy, crinkly leaf that is "fire-cured" by hanging it in barns full of thick smoke. The curing barns are small and as airtight as possible. Few of the "fire-cured" barns are truly airtight, and smoke often pours from them during the fall curing season. A visitor could be excused for thinking that most of the tobacco barns were on fire. Farmers lay sticks of green hardwood on their barn floors after they have hung their tobacco crops, and they set smoldering fires to cure the leaf. They may sprinkle damp sawdust on the smoldering wood to produce even more smoke. The leaf acquires a distinctive creosotelike flavor from the fire-curing process. For some strange reason it has not been in great demand in recent years, although it is used in making snuff.

Far and away the most famous of the limestone lowlands in the Bluegrass area of north-central Kentucky, which owes its reputation to 200 horse farms in a 15-mile semicircle north of Lexington. The Bluegrass area is a gently undulating plain underlain by limestone rich in phosphorus. The soils derived from this limestone are remarkably fertile, and they are two to four feet deep.

The Bluegrass horse country is the most beautiful rural area in the United States, if not the entire world. Brood

Figure 5–1. The Interior Low Plateau and the Limestone Lowlands. The Interior Low Plateau of Kentucky, Tennessee, and adjacent states has four productive limestone lowlands. The Bluegrass basin of Kentucky and the Nashville basin of Tennessee are low geological domes that have been "scalped" by erosion. The Pennyroyal Plain along the Kentucky-Tennessee line and the limestone valleys of northern Alabama are geological "saddles" on the upland surface.

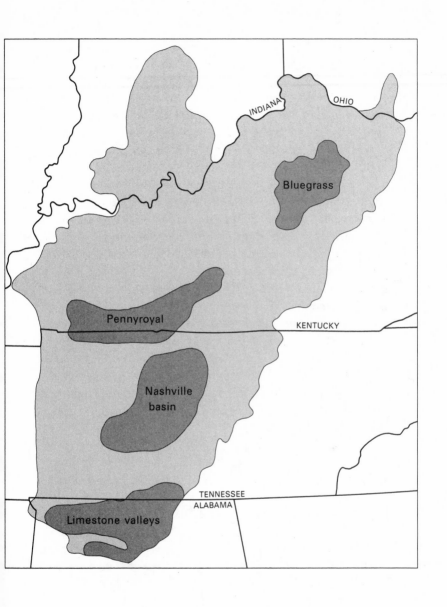

mares graze placidly, and long-legged foals cavort in neatly manicured pastures of succulent bluegrass. Dazzling white board fences follow the gentle roll and swell of the countryside. The board fences are practical as well as pretty because the high-spirited horses could easily injure themselves by running headlong into a wire fence.

Enormous old oak, walnut, and sycamore trees dot the pastures and march along country lanes that run between walls of unmortared fieldstone, now beautifully weathered and moss-grown. Trees guard the long driveways that lead back to the elegant, white-pillared mansions, and they shade the luxuriant mansion grounds. The paneled horse barns behind the mansions cost more than most suburban homes, and many mansions have their own private swimming pools and tennis courts. The whole countryside has a quiet air of contented wealth.

Americans first began poking into the Bluegrass area around 1766. The early explorers were adventurous "long hunters," of whom Daniel Boone has received the most publicity, and land speculators, who staked out large claims. It has become fashionable to think that land speculation was bad, but in those days making a fortune on frontier land was no more wicked and sinful than making a fortune in the stock market is today. George Washington himself was not above speculating in western lands, and he might have been even more avid if he had ever laid eyes on the Bluegrass.

The Bluegrass country was the first area west of the Appalachian barrier that Americans settled and brought into agricultural production. Their staging area was the Great Valley of East Tennessee, their gateway was the Cumberland Gap, and their route was the Wilderness Road blazed by Daniel Boone in 1775. The Wilderness Road actually was little more than a trail for packhorses, but it brought a steady stream of settlers from Virginia, North Carolina, and Tennessee. They drove herds of cattle, hogs, and sheep before them, and they brought horses and mules as pack and riding animals.

Kentucky had no permanent Indian settlements, but it was a prime hunting area for Shawnees from north of the Ohio River and Cherokees and Creeks from the south. The early white settlements were frequently attacked by Indians, who wanted to preserve their hunting grounds, and many settlements were abandoned more than once. The last major Indian "invasion" was ended at Blue Licks in 1782, the final battle of the Revolutionary War, but sporadic raids continued until 1794, when the battle of Fallen Timbers effectively ended Indian influence in Ohio.

Kentucky was as fertile for lawyers as it was for crops because much of the land was settled before it had been surveyed. It became a hodgepodge of conflicting land claims that engendered a veritable flood of litigation. The state of Virginia, of which Kentucky had been a part until it was admitted to statehood in 1792, had paid some of its war veterans with warrants that entitled them to 50 to 5,000 acres of land in Kentucky, depending on their rank and length of service. The warrants did not specify the location of this land, however, so overlapping claims were virtually inevitable, and the claimants with the best lawyers got the best land.

Men of wealth or influence were able to assemble large Bluegrass estates, while smaller farmers were pushed back into the poorer, hilly areas or had to move on to the newer lands that were opening in the West. Many of the large estates were acquired by the well-to-do younger sons of prosperous old Virginia families. They moved in with their black slaves, transplanted their farming system and their entire lifestyle, and became the cultural and political aristocracy of the state. By 1800 they had already started to build fine mansions in which they could enjoy the lives of country gentlemen.

From the very start most Bluegrass planters and farmers turned west and south rather than east in search of a market for their products. They could drive herds of cattle, hogs, sheep, mules, and horses over the Wilderness Road to eastern markets, but packhorses were the chief means of transport eastward, and they could only carry small loads.

Transport downriver was much easier, and most farmers looked to New Orleans for their principal market. In the fall, when their crops had been harvested, some farmers would cut down trees, build their own flatboats, load them with the products of their farms, and float them downstream to New Orleans for sale and shipment to Europe or to the East Coast. After they had sold their goods they broke up their flatboats, sold the wood for building material or firewood, and made the long trip home overland.

The Kentucky riverboat men were a rough lot, who liked to boast that they were "half horse and half alligator," and the good people of New Orleans were only too happy to see them depart. The lucky ones could afford to buy horses, but many had to walk back to Kentucky. Their route took them northward to Natchez, and then across country to Nashville by way of the famous Natchez Trace, which was proclaimed a post road as early as 1800.

Some Yankee "pack peddlers" made a good living from early Kentuckians by traveling a circular route. They brought packhorses loaded with goods from the east through Cumberland Gap and over the Wilderness Road to the Bluegrass, where they traded their goods for tobacco. They carried the tobacco to Louisville, sold their horses, and loaded the tobacco on flatboats for New Orleans. In New Orleans they sold the tobacco and took passage around to the East Coast, where they started the same trip all over again.

The principal products of Bluegrass farms have included hemp, corn, tobacco, and livestock. Hemp (L. *Cannabis*), no longer a commercial or legal crop in the area, is a tall, graceful plant that has long, coarse, stringy fibers. These fibers could be made into tough rope and durable duck sailcloth. They could also be made into burlap, which was shipped down the river to cotton gins for wrapping bales of cotton after they had been ginned. In the early days no one smoked the leaves.

During the nineteenth century hemp faced increasing competition from jute and manila fibers produced in southeastern Asia, and Bluegrass farmers virtually stopped growing the crop. They responded patriotically to government pleas to start growing it again during World War II, when the supply of fiber from Asia was interrupted. They stopped again after the war, and hemp is no longer grown commercially, but despite the best efforts of law enforcement officers it still grows wild along roads and fencerows, perhaps with some nocturnal encouragement.

The very first settlers in the Bluegrass planted corn for themselves and their animals, but almost from the start they distilled part of the crop and stored it in jugs and barrels. Water from limestone springs gave them a superior product, and they discovered that it could be improved even more by storing it for a while in white oak barrels with charred interiors. The charred barrels absorbed foreign particles, mellowed the sharp taste, and gave the whiskey a pleasant amber color. By 1810 Kentucky boasted no fewer than 2,000 distilleries, and a decade later the state was shipping 200,000 gallons of whiskey a month to New Orleans alone. Bourbon whiskey took its name from Bourbon County, which originally included much of northeastern Kentucky. The county was named for the French royal family, some of whose lesser members took refuge in Kentucky after the French Revolution.

Today the state has 28 distilleries, mostly in Louisville or in the belt of hills to the south. They use a billion tons of corn, rye, and barley each year to produce two million barrels of bourbon worth $300 a barrel. The distilleries sell their spent mash to dairy farmers as feed for their cows, and these farmers must have some of the most contented animals alive. The whiskey is aged from four to twelve years in barrels that are stored in enormous bonded warehouses under the close scrutiny of the Internal Revenue Service, which collects about a billion dollars a year from the distilleries in excise taxes.

The third important crop in the Bluegrass triad is Burley tobacco, which is the only significant cash crop. Tobacco makes so much money that even the showplace horse farms grow as much as they are allotted, five to ten acres or even more, but the tobacco field is tucked away in a remote part of the farm where it is less likely to be noticed. Nevertheless, many tobacco farmers in the Bluegrass area paint the vertical ventilator doors of their curing barns in a color that contrasts with the body of the barn, giving them a distinctive striped appearance.

Despite the importance of bourbon production and tobacco farming, livestock rather than crops have dominated the Bluegrass economy from the very first. Early herds flourished, and the settlers decided that something in the grass and water gave their animals exceptionally strong bones. By 1800 large landholders were importing fine breeding stock from England. They specialized in producing superior animals that could be sold as breeding stock to farmers in other parts of the United States. All of the best families were involved. The first major livestock show was organized in Lexington in 1816, and the shows were as much social and political as agricultural events.

Livestock were an excellent product for Bluegrass farms in the early days when overland transportation was poor because the animals could walk to market. Herds of fat cattle and hogs from the Bluegrass were driven eastward over the Wilderness Road to cities on the eastern seaboard, and the spread of cotton production in the South created a major market for workhorses and mules bred in the Bluegrass. Isaac Shelby, Kentucky's first governor, drove herds of mules to cotton plantations in South Carolina, and once he corralled his animals at the governor's mansion in Columbia when he spent the night there.

Beef cattle account for about 20 percent of the total income of farms in the Bluegrass, but almost from the start the breeding of horses and mules has been more important. It is not easy for a contemporary American to appreciate the numbers of horses and mules needed before the days of the tractor, the truck, and the automobile. Farmers had to have horses and mules to work their land, city merchants needed dray horses to pull their wagons, and the ownership of fine carriage horses was a socially acceptable way of demonstrating wealth and status. Nineteenth-century Bluegrass breeders specialized in raising fine horses for carriages, for riding, and of course, for racing.

Kentuckians have always had a passion for horse racing. It was a popular pastime on the frontier, and men plotted, schemed, and taxed their wits trying to figure ways to breed faster horses than their neighbors. The first, impromptu races were held in the streets of towns, which provided the best available straight stretches. The streets had the additional advantages of making available a ready audience and convenient taverns for celebrations afterwards. As early as 1787 the town trustees in Lexington banned horse races on Main Street, and a formal race course was laid out at the edge of town in 1789. By the

1830s the Bluegrass animals had begun to compete on the turf with animals from older breeding centers in the East, and these races generated such excitement that half the population of Louisville turned out to watch Grey Eagle run against Wagner in 1839.

It was quite proper for a gentleman to breed fast horses, but caring for them, training them, and riding them in races was considered beneath his dignity. In the nineteenth century most grooms, trainers, and jockeys were black, first slaves and later freedmen. The slaves were housed in a hamlet at the back of the estate, away from the mansion. After the Civil War the slave hamlet was turned into a freetown, and each freedman was given a house on a lot large enough for a garden, a pigpen, and a chicken coop. The owner thus ensured himself of a cheap labor force. The Bluegrass area still has a number of such hamlets, some of which are now inhabited by white farm workers rather than by blacks.

After the Civil War horse breeding in Kentucky received an enormous boost when wealthy northerners began flocking to the Bluegrass to buy horse farms and the status they conferred. Horse racing, after all, is "the king of sports and the sport of kings," and the fine horse has always given special status to its owner. The man on horseback, the knight, the conquering warrior-hero, has been a commanding figure throughout European history. The landed gentry in England rode to hounds and loved horse races, and this tradition has been transplanted to the upper echelons of American society.

Horse farming in the Bluegrass is a billion-dollar business. It is dominated by the breeding of Thoroughbreds, which are race horses. The breeding, training, and racing of Thoroughbreds are separate operations. Some breeding farms train and race their own animals, but many of the

Thoroughbred horses raised on Bluegrass farms are sold, and some farms send their animals to specialized training establishments.

Horse farms confer social status as much as economic advantage. Undoubtedly some horse farms do manage to make money on a more or less regular basis, but a horse farm is an enormous gamble. The cost of operating even a small horse farm can run well over a million dollars a year, and the owner must be prepared to spend a thousand dollars as casually as most people would spend a dime. The owner may be able to write off his losses on his income taxes, but as J. P. Morgan said when he was asked about the cost of a yacht, "If you have to ask what it costs, you can't afford it!"

Woodvale Farm is south of the Georgetown–Paris Pike just east of the former slave village of Centerville. The main house is a story-and-and-a-half structure of white-painted brick. It looks small, but it has four bedrooms, and the three-car garage behind it has five bedrooms upstairs for domestic help. The house sits on a tree-shaded, eight-acre lot that overlooks the rest of the farm. The entrance to the paved drive leading to the main house has a Bluegrass "lazy gate" that can be opened or closed by pulling on a rope attached to a set of levers, without having to dismount from a horse or get out of a car. In 1959 the gate was chained shut and padlocked, and it was easy to get the impression that no one lived there.

Just east of the main house the view was like a picture on a calendar. A paved farm road curved down to a creek and then up to the farm manager's house, an old log structure long since clapboarded over, which was set in a clump of trees. The road was framed by white-painted board fences, and the grass strips between the road and the fences on either side were planted with shade trees. White board fences divided the farm into eight separate pastures, and they enclosed eight smaller paddocks of an acre or so each. (Figure 5–2)

Forest Mynear, the farm manager, told me that he started working on a farm for $9 a month and board when he was 13 years old, and got what little schooling he could on the side. "I came here when I was 21 years old, and I've been here for 41 years. This has been a horse farm ever since I've been here. I get a salary plus a percentage of everything sold from the farm. I've worked for three different owners. In 1948 the farm was sold to Mr. W. Alton Jones, who also owns an Angus cattle farm in Maryland, a quail plantation in Georgia, and a place in Florida. He lives in New York City."

W. Alton Jones was the youngest of seven children who were raised on a rocky, 40-acre farm near Joplin, Missouri. He worked days in a grocery store and nights in a bookstore to earn enough money to go to college, but he had to drop out at the end of his freshman year because of his mother's sudden illness. He went to work as a $45-a-month janitor and meter reader for a utility company, and in his spare time he took a correspondence course in bookkeeping. The utility company became the Cities Service Company, a multibillion dollar oil company. In 1939, when he was 48 years old, Mr. Jones became president of the company, and in 1953 he was made chairman of the board. He was one of the nation's highest paid executives. Mr. Jones

Figure 5–2. Stonereath (formerly Woodvale) Farm. The entire farm is enclosed by painted board fences, and the lush green pastures are dotted with clumps of shade trees. The main house, garage with servants' quarters, and swimming pool are in the center north. The shaded rectangles are horse barns, and the solid rectangles are rows of houses for workers on the farm. Reproduced by courtesy of Darrell Brown.

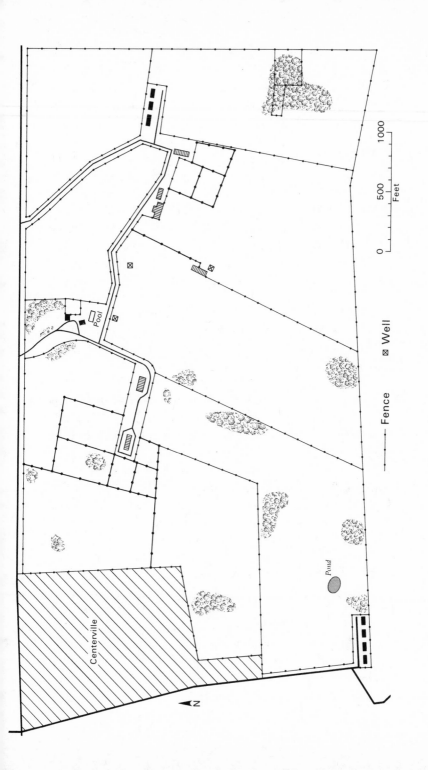

Centerville

Pond

Pool

N

—•— Fence   ⊠ Well

0   500   1000
Feet

was a good friend of President Dwight D. Eisenhower. In September, 1959, he flew from New York to Scotland just for a weekend of golf and bridge with the president. On January 16, 1960, the *New York Times* had a front-page photograph of Mr. Jones welcoming the president to his 14,000-acre quail-hunting plantation near Albany, Georgia. On March 1, 1962, he was killed when his plane crashed on takeoff en route from New York to Mexico for a fishing trip with Eisenhower. In his wallet the police found a $10,000 bill, a $5,000 bill, a $1,000 bill, several $500 bills, and some small change. In his briefcase he had $45,000 more, just in case he happened to want a cup of coffee or a Cadillac.

In 1959 Forest told me, "I don't see Mr. Jones but twice a year, sometimes not then. He always comes for the Derby and sometimes for the fall races, too. He and Mrs. Jones fly to Lexington in their own plane. They have a servant drive a car from New York to open the house and meet them at the airport. They throw big parties during the races." Later one of the workers said to me, a bit wryly, "Mrs. Jones just loves this place. She once spent a whole week here."

Forest obviously ran the farm pretty much as he saw fit, and he acted as though he owned it. It was primarily a horse farm, but it also had a substantial herd of beef cattle and a sizable tobacco base. Five gleaming white horse barns had stall space for 72 horses, but there were only 20 mares on the place in 1959. "This is the year we had to sell some for tax purposes," Forest said. "We sell a lot and make a big profit to keep the Internal Revenue people happy, and then next year we buy them all back and write off the cost as a business expense.

"We breed the mares to foal as soon as possible after the first of January because that's the official birthday of every Thoroughbred. The colts are weaned in September, and

we start breaking them at the end of the year. The next fall
they go for training, and they start racing when they're
two-year-olds. The only race that really counts is the Kentucky Derby. A few years ago one of our horses was fourth
in the Derby, and another horse from this farm has also
placed in it.

"In summer we keep the horses in the barn during the
day to protect them from the sun and the flies, and they go
out to graze at night. Two men are kept busy full time just
taking care of the horses, cleaning out the manure and
putting fresh straw in daily. I buy all of the feed for the
horses from the feed store. I also have to buy hay because
we don't make any."

The farm had a herd of 65 Angus beef cattle, and Forest
had four milk cows and eight sows of his own. "I have 12
acres of corn, 6 and 52-hundredths acres of tobacco, and
the rest is pasture. I raise corn for the hogs, and sell the
tobacco. The farm runs around 300 acres, and there ain't
five acres on it you couldn't cultivate if you wanted to, but
some of the pastures have not been plowed for 25 years.
They get fertilizer and up to two tons of magnesian limestone every six to eight years. We plant trees in the pasture
to make shade for the horses.

"We mow the pastures three times a summer to keep
down the weeds," Forest told me. "It takes two men with
two tractors fourteen days to clip the entire farm one time,
but mowing grass is only our second biggest job. Painting
fence is the biggest. We paint so much fence every year
and get around the entire farm every three or four years.
We buy paint in 55-gallon drums, and store it and mix it in
a paint shed."

In 1959 Woodvale had nine full-time employees. Two
tended the horses, two clipped the pastures, and one was
the night watchman. "B. O. Earlywine, he don't do nothing but mow the lawn up at the main house and the grass

strips along the paved farm drives with that little ol' riding power mower, and sometimes we have to go up and help him out when he's got his hands full." The other three workers could be called on for whatever had to be done, but as far as I could tell their main job was to paint fences. (One of the workers muttered to me, "Them people in that big house up there think more of them damned horses than they do of us.")

All nine of the men lived in Centerville, half a mile east of the main house, in houses provided rent-free by the farm. "Centerville used to be a colored settlement," said Forest, "but the whites like to run 'em all out, and they done moved in to Lexington. Just about everybody in Centerville now is white and works on one of the farms around here."

Centerville had not changed very much when I went back for a visit in the spring of 1983, but Woodvale had been transformed. The fences were all painted black, the barns and other outbuildings were cream, there were no crops in sight, and the entire farm looked better than ever. The entrance to the main driveway was not locked, and the house was occupied. The new owners, Darrell and Lendy Brown, had changed the name of the farm to Stonereath, a combination of Lendy's maiden name, Firestone, and her mother's maiden name, Galbreath (See Figure 5–2).

Darrell Brown, 49, grew up in Oklahoma and has a degree in accounting from the University of Oklahoma. He learned to fly jet fighters in the air force, then went to work for the Kerr-McGee Oil Company and rose to senior accountant, but stayed in the Air Force Reserve. He demonstrated private jet airplanes, and after Arnold Palmer bought one, he went to work for Palmer for seven years.

"I've always been around horses," Darrell said. "We had

quarter horses in Oklahoma, and Arnold had horses. Thoroughbreds are the top if you are in horses. We had been in Ocala, but Lendy had bought the farm here, and we moved here eight years ago. The Bluegrass is the only place to be if you are in the horse business. It has the best stallions, the best buyers, the best prices, the very best of everything."

Darrell told me that the farm had been leased out after Mr. Jones died, and it was pretty run down when they came here. They are developing it into a top-notch horse-breeding farm. They have gotten rid of the cattle and the other sideline activities. "We still have a 60,000-pound tobacco base," Darrell said, "but we only have barn space for 24,000 pounds, so we lease out the rest.

"We changed to black fences four years ago because they're more economical. We let the white fences go for a year, and they looked pretty bad. Then we scraped them, and we've repainted them three times. Now we'll only need to repaint them every other year. We hire a contractor with a machine that sprays both sides simultaneously.

"Four years ago we bought a second farm, three miles down the road toward Paris. We paid $2,000 an acre for 500 acres. Now they're asking $6,500 an acre for a 150-acre farm across the road from it. The farm we bought was one of the better farms in the area. The farmhouse was built in 1815. We have converted it into an office. We tore out all the old wire fences and put in new board fences, and we paved all the roads. We rebuilt the old tobacco barns as horse barns and built three new horse barns. We fenced off nine small paddocks beside the barn in front of the office. The paddocks are for colts and fillies getting ready for sale—you don't want them getting all banged up. They are also useful for animals that are hard to catch, or for sick animals, or for any that have to have special treatment. You never have enough paddocks. We are still building up the place, and this year we are starting on another new 30-stall barn. Two years ago we bought another farm of 200 acres at the back of this one."

They buy all of their feed and use the land only for grazing. They want to get the whole farm into permanent pasture. One man with a batwing mower is constantly clipping pastures, and they may need a second man. He goes over the entire farm every week or ten days, and once he has finished he goes back and starts all over again.

"We employ thirty people full-time. We used to have a resident manager on each farm, but now Pete Cline is in charge of the entire operation. We furnish him a residence on the new farm and an automobile in addition to his salary. We've also built four new houses for key people at the southwest corner of the old place. We used to furnish free housing and utilities, but we stopped that after one guy showed them his pay check and got food stamps. It is better to pay people more and let them worry about their own houses."

They have 130 mares on the two farms. The foaling mares and any horses that need special attention are at the old place. They wean the foals in the fall and take them to the new place. The new place has the stallion operation, 45 yearlings, and all of the maiden and barren mares. They have two stallions, and are adding one next year. The stud fees are $15,000 for Grand Zar and $10,000 for Joanie's Chief.

"Our own mare, Best in Show, is worth $4 million to $5 million," Darrell told me, "but our best horse is a filly named Blush with Pride that earned $536,000 last year. We have syndicated Gold Cup, a weanling by Alydar out of Best in Show, for $2,500,000. The members of the syndicate share its expenses, and they share everything it wins. We sell yearlings at Keeneland, which has four thoroughbred sales each year. The select sale for yearlings is in July.

Four thousand yearlings are nominated for the select sale
and only 350 are selected. They check the pedigrees, and
then they come out and look at the horses."

The Keeneland Selected Yearling Sale made headlines
all over the country in July, 1983, when an Arab sheikh
paid $10.2 million for a single yearling colt, and 301 year-
lings were sold for a total of $150,950,000, or an average of
$501,495. As J. P. Morgan said, . . . !

6

The Second
Seedbed

The agricultural heartland of North America is the Corn Belt, a vast expanse of corn and soybeans that stretches for 800 miles across the rolling plains of the Middle West from central Ohio to eastern Nebraska. The seedbed and cradle of the Corn Belt was southwestern Ohio, where three major streams of migrants converged to create the first truly American people (Figure 6–1).

The early settlers of southwestern Ohio were not insensitive to their origins, and many went to their graves thinking of themselves as Yankees, Pennsylvanians, Virginians,

Figure 6–1. Westward movement of occupance systems in the eastern United States. The St. Lawrence stream moved westward through the boreal forest. The Yankee and Yorker stream of wheat-then-dairy farmers moved to the "backbone counties" of northeastern Ohio, thence to the upper Lake states. The northern branch of the Middle Colony stream from southeastern Pennsylvania moved west by the National Road or the Ohio River to southwestern Ohio, seedbed of the Corn Belt. The Appalachian branch of the Middle Colony stream followed the Great Valley to the Watauga country of northeastern Tennessee, and then moved through the Cumberland Gap to the Bluegrass and then on to southwestern Ohio. The tobacco plantation stream moved from the Tidewater of Virginia to the Kentucky Bluegrass country. The plantation stream from the Carolina coast changed from rice to indigo, then to cotton, as it moved inland onto the Piedmont. The sugar plantation stream moved up the Mississippi River from New Orleans, but it could not compete with the cotton plantation stream from the east.

or even English, Scotch-Irish, or German. Their offspring
were basically a new breed, however, an alloy blended of
distinctive contributions from many different parts of the
eastern seaboard. The influence of Europe was much
weaker than it had been in the East because Americans

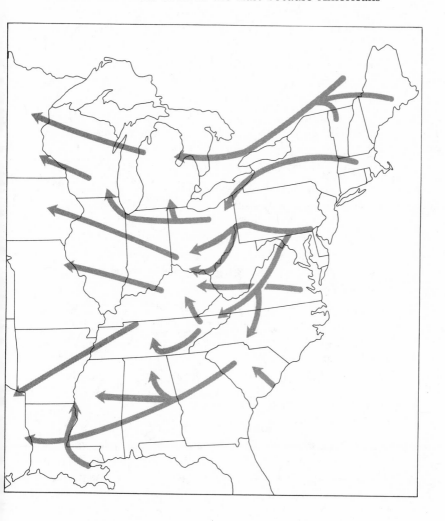

had learned how to cope with the wilderness, and immigrants who came directly from the Old World were happy to benefit from their experience.

Three streams of migrants fed southwestern Ohio. The southern stream came overland through the Cumberland Gap, over the Wilderness Road, and across the Bluegrass to the Ohio River. The second stream, which was fed by several branches from the east, took the river route and floated down the Ohio from its fork at Pittsburgh. The shortest and most direct route westward for people from Pennsylvania and the other middle states was also the most arduous because it crossed the difficult terrain of the Appalachian Uplands before it reached the Forks of the Ohio at Pittsburgh. When the pioneers finally reached the river that flowed west, they built or bought flatboats, piled all of their worldly goods on board, and floated downstream in search of a new home. The flatboats were ungainly craft, rectangular boxes 15 feet wide and 40 feet long, with a roof at one end for the family and a pen at the other for their animals. The jampacked flatboats were a weird combination of log cabin, floating barnyard, country store, and fort.

The third stream of migrants to southwestern Ohio, the last and probably the largest, came overland by the National Road, which did not reach Springfield until 1838. The National Road was a response to the needs of the newly emergent nation. The Ohio River was essentially a one-way route, especially for bulky goods such as farm products. It was easy, if hazardous, to drift downstream, but travel upstream against the current was extraordinarily difficult until the steamboat era, which did not begin until after the War of 1812. The Appalachian Uplands were a further barrier between the head of navigation on the Ohio and the markets of the East.

The settlers in the Ohio Valley had to look downstream

toward New Orleans for a commercial outlet, and periodically they muttered about seceding from the United States and forming a new and independent nation west of the Appalachian Uplands. The people in the Ohio country were veterans of a war of independence from Britain, and the idea of another war of independence did not seem nearly so alien to them then as it does to us today.

Congress was sensitive to the mood of the West, and in 1806 it authorized the construction of the National Road, a wagon road westward from Cumberland, Maryland. The original plan called for the National Road to run all the way to St. Louis, but Congress got tired of paying for it by the time it had reached Vandalia, Illinois. Today it is paralleled by U.S. 40, and west of the Ohio River by Interstate 70. The capitals of Ohio and Indiana have blossomed where the National Road crossed major rivers: Columbus on the Scioto and Indianapolis on the White.

The National Road was the first improved highway between the middle states and the Midwest. It was good enough for farm wagons, and it became the major route of migration from Pennsylvania and neighboring states to Ohio, Indiana, and Illinois. People from the middle states moved westward along the National Road, and then spread out in both directions between the Yankees and Yorkers to the north, who were moving westward along the Erie Canal and the Great Lakes, and the Virginians and Kentuckians to the south, who followed the valleys of the Ohio and Mississippi rivers and their tributaries.

The pioneers floating down the Ohio River found their first easy access to the fertile plains of the Midwest by way of the Miami and Little Miami river valleys in southwestern Ohio. For the first 400 miles downstream from Pittsburgh the Ohio River flows through the hills of the Appalachian Plateau, which are unrewarding country if you are thinking about making a living by farming. Narrow strips of productive bottomland border the river, first on one side and then on the other, but the bottomland is rarely more than a mile wide and behind it are hills too steep for cultivation.

The river takes leave of the plateau near Maysville, Kentucky, but its bottomland is still boxed in by belts of hills some 10 to 15 miles wide that have been fretted by its tributaries. The valleys of the Miami rivers were the first easy passage through the hills to level lands that stretched off to the north and west forever, or so it seemed. Furthermore, beyond the Miami River the Ohio turns southward, the wrong direction for settlers headed for the interior. Cincinnati, founded in 1789 between the two Miami rivers and outside the great bend of the Ohio, was the best place for pioneers to leave the river and set out for the interior. It became the gateway to the Corn Belt.

Cincinnati was the outfitting point where settlers could equip themselves with the tools and implements they would need on the frontier, and it was their contact point with the rest of the world. It was the collecting point to which they brought their products when their land started to produce a surplus, the processing point where their products were prepared for shipment, and the distribution point at which they could buy the goods they could not produce on their own land.

Settlers came to Ohio from three major source areas that had already developed distinctive systems of farming, and their early efforts on the frontier reflected their backgrounds and experiences. The Yankees and Yorkers who settled northeastern Ohio had come from areas where wheat was the most important crop. They grew wheat as a cash crop, and shipped their wheat and flour eastward along the Erie Canal. The southerners in southwestern Ohio had come from areas where corn and tobacco had been the principal crops, and they continued to grow them. The Pennsylvanians in southwestern Ohio came from areas of mixed farming. They had raised crops to fatten cattle and hogs, they had returned the manure from their animals to the soil to enrich it for crop production, and they understood the value of crop rotation. They transplanted the farming system of southeastern Pennsylvania to the new seedbed in southwestern Ohio so successfully that the upper Miami River valley west of Dayton even looks a bit like Lancaster County.

Some aspects of the Pennsylvania farming system were modified in southwestern Ohio because members of other groups had learned better ways of doing some things. This modified farming system eventually spread across the Midwest from its seedbed in southwestern Ohio until it reached its climatic limit in eastern Nebraska. It remained the norm in the Corn Belt until after World War II and turned the Midwest into one of the human wonders of the world.

The Corn Belt farming system was based on a rotation of corn, a small grain (either winter wheat or oats), and a leguminous hay (clover in the early days, later alfalfa). For more than a century these five crops—corn, wheat, oats, clover, and alfalfa—were virtually the only field crops grown in the Corn Belt. In 1924 the first modern census of agriculture found that the five crops of the traditional rotation accounted for more than 95 percent of the harvested cropland in the four Corn Belt states of Iowa, Illinois, Indiana, and Ohio. Corn Belt farmers used these crops primarily to fatten hogs and cattle, although they sold their wheat for cash.

Corn, the linchpin of the Corn Belt farming system and the backbone of American agriculture, is a plant of subtropical origin. Unlike the small grains, corn is physiologically capable of taking full advantage of the long growing season and abundant precipitation of the Corn Belt. Plant breeders have developed varieties of corn that have greatly extended the area in which the plant can be grown successfully, but traditionally it has required long hot summers with scorching days and sultry nights. The optimum has been a growing season of 140 days or more, an average summer temperature of 75° F, and abundant rainfall during the growing season, especially during the critical tasseling period in early summer.

Corn was an excellent crop for the pioneer farmer. It gave him more grain per acre than wheat, and he could plant it between the stumps on newly cleared land, whereas wheat required better soil preparation. He normally tried to plant his corn at the end of April. An old jingle warned him that the cost of planting late was a bushel a day after the fifteenth of May. Corn was less susceptible to diseases and insects than wheat, but the farmer had to protect his growing crop against crows, blackbirds, squirrels, and a host of other varmints. He had to cultivate it with hoe or plow to get rid of grasses and weeds until it was knee high and could be "laid by" until harvest time. Another well-known jingle identified the customary goal of a crop that was knee high by the Fourth of July, so the farmer could lay by his cornfield and get on about his other business.

Early farmers harvested their corn in a variety of ways. Some simply left the stalks standing through the winter and snapped off the ears only when they needed them. Some picked all the ears in the fall, stored them in a crib, and turned cattle in to the field to graze the stalks. A farmer might "hog down" the entire crop if the field was securely fenced and he did not plan to plant oats on it until the following spring. If he planned to plant winter wheat, which had to be sown as soon as possible after the corn had ripened, he cut the stalks, stacked them in shocks, and sowed the wheat around them.

A small grain, either wheat or oats, followed corn in the traditional rotation. In the southern Corn Belt corn matured in time for the farmer to plant winter wheat. Wheat was a risky crop for farmers farther north, however. They waited until the following April and planted oats rather than wheat when it was still too cool to plant corn. The farmer harvested his small grains in July after he had laid by his corn. The small grains helped to spread his work load because they required labor when he was not busy with corn. He normally sold wheat as a low-profit cash crop, but oats rarely were worth the cost of hauling them, so he fed them to his own livestock. Oats were especially useful for feeding workhorses in the days before tractors. Both wheat and oats also provided straw for bedding and roughage, but perhaps their most useful contribution was to serve as a "nurse crop" for the hay crop that followed them in the traditional rotation.

Both clover and alfalfa, the two principal leguminous hay crops of the Corn Belt, are delicate when they are young, and they may not grow well if they are planted on bare ground. The farmer normally sowed them in April in his young wheat field or with his oats. The faster-growing grain protected the young forage plants against sun and driving rain, and they would have become well established by the time the small grain was ready to be harvested.

Clover was the principal hay crop of the Corn Belt until a couple of generations ago because most farmers thought that alfalfa would not grow well in humid areas. Both alfalfa and clover are perennial plants that will produce new growth for several years without having to be reseeded, but clover is a weak perennial. The farmer usually planted timothy and other grasses with his clover. In the first year his hay was mostly clover, but his second year's crop was about half grass, and if he left the field for more than two years it became mostly grass, which does not make as good hay as clover does.

Corn Belt farmers almost completely abandoned clover once they had learned how to grow alfalfa, which produces more tons of hay per acre, and hay with a greater feed value per ton, than any other perennial feed crop. Alfalfa puts out lush early growth that helps to shade out troublesome weeds, it grows back rapidly after it has been cut, it has a deep and extensive root system that helps it to withstand dry periods, and, like clover, it is a legume that extracts nitrogen from the air and stores it in the soil where other crops can use it.

After a year or two the farmer would plow up his hayfield and plant it with corn, which thrives on the nitrogen stored in the soil and on the organic matter (humus) that had been added to the soil when the hay plants were plowed under. The farmer also would give his cornfield all of his barnyard manure and any fertilizer that he used because corn was his most important crop. The small grain had to make do, but did quite nicely, with the residual fertility left over from the corn crop.

The basic crop rotation of the Corn Belt persisted for more than a century because it was economically successful and ecologically sound. It leveled off peaks and troughs in the distribution of labor and income, it facilitated control of weeds, pests, and diseases, and it maintained and even increased the fertility of the soil. The labor requirements of the different crops dovetailed nicely to balance out the seasonal distribution of labor and to keep the farmer productively occupied for much of the year. He was spared the grief of sharp fluctuations in income because he did not have all of his eggs in one basket. He derived income from several crops, and a poor year for one crop would probably be a good year for another.

Growing different crops in succession prevented the

buildup of pests and diseases in the host plants they favored, and it retarded the spread of new pests and diseases. A new disease or pest can spread like measles in kindergarten in an area that concentrates on a single crop. Most weeds are annual plants that grow from seeds, and the farmer can control them by killing the plants, either by plowing them under in a cornfield or by cutting them off in a hayfield, before their seeds have had a chance to develop.

Different crops remove different amounts of different nutrients from the soil, but nutrient withdrawal is balanced over the course of a rotation. The alternation of deep and shallow rooting crops equalizes nutrient withdrawal from the different soil horizons. The leguminous hay crop is the principal soil builder. It transfers nitrogen from the atmosphere to the soil, it adds humus to the soil when it is plowed under, and it helps to reduce erosion that the cultivation of row crops, such as corn, encourages. Cultivation in rows helps to control weeds, but the bare ground between them is easily eroded.

A knowledge of the basic rotation is the key to understanding what crops were grown, and why, but no farmer was bound by it, and it was modified to fit the particular conditions of particular fields and farms. On fertile level land, for example, a farmer might take two crops of corn before he planted small grain, and farmers in some unusually favored areas grew "continuous corn," with no rotation whatsoever. At the other extreme, on steep land subject to erosion, the hay crop might be left a second year or even longer and used as a rotation pasture. Grass and hay crops in a rotation are like the brakes on an automobile—the steeper the slope, the more you need.

Farmers around the edges of the Corn Belt modified the standard rotation in response to increasing environmental stress. To the west, for example, the rainfall is so light and so undependable that farmers could not be sure that they could make crops of corn or hay, so they concentrated their efforts on the small grain, wheat. Farmers in the hilly areas to the south and east practiced the standard rotation on level areas, but much of the land was too steep to be used for anything except pasture or woodland.

Farmers north of the Corn Belt had to cope with a short, cool growing season and difficult terrain. They had to worry that an early frost might ambush their corn before the kernels had a chance to ripen, so many of them harvested the crop for silage rather than for grain. They cut the entire plant while it was still green, chopped it into fine pieces, and stored it in cylindrical silos as winter feed for their livestock. The youthful glacial terrain north of the Corn Belt is choppy, with many short, steep slopes. Good management required that much of the land be used for pasture rather than for crops, and that much of the cropland be used for hay rather than for row crops. Furthermore, the farmer in the north, where winters are longer and harder, had to put up more winter feed than farmers farther south, where animals could be kept on pasture for a greater part of the year.

The early Corn Belt farmers in southwestern Ohio had to figure out what to do with their crops once they had grown them. Everyone used the same rotation, everyone raised the same crops, everyone produced a surplus, but no one had a way to sell it. The cost of transporting crops eastward, up the river and across the Appalachian Uplands, was prohibitive until railroads had been constructed, and even shipping them down river was expensive.

The best solution, the farmers realized, was to convert their bulky crops into commodities of lighter weight and greater value that could bear the cost of transportation to distant markets. They could mill their wheat into flour and distill their corn into whiskey, but the best way to dispose of their crops was to "march 'em off to market" on the hoof as fat cattle and hogs.

Farmers could usually make more money per acre with hogs than with cattle because hogs have a much faster turnover rate. A cow has only one calf a year, but a sow can have a litter of six to eight pigs every six months. Today it takes a year to a year and a half to feed a calf to market weight, and in the old days it took even longer. Hogs eat like hogs. They can be fattened and ready for sale in six months or so from the time they are born. Like people, hogs have shorter intestines, so they can use concentrated feed, like corn, more efficiently than cattle can (Figure 6-2). It takes four pounds of corn to make a pound of pork, but eight pounds of corn and ten pounds of hay to make a pound of beef.

Cattle have advantages, however. They may need more land than hogs, but they can make better use of poor land. They must have roughages as part of their regular diet, but their complex digestive systems mean that they can convert roughages such as pasture, hay, and cornstalks into meat and milk. Cows require less labor than hogs.

As a general rule, smaller farms are more likely to have hogs because they have the labor and they need the income, whereas larger farms are more likely to have cattle. Many a hog farmer has dreamed of the day when he could "trade up" to beef cattle, which require less labor and confer greater prestige. The hog farmer does not have the glamour of the cattleman, although most hog farmers take quiet pride in doing a good job.

Southwestern Ohio has always had the smallest farms in the Corn Belt. The farmers have fattened hogs rather than cattle, and fat hogs have been the principal product of the small farms in the eastern Corn Belt for a century and a half. The earliest Corn Belt farmers' hogs were razorbacks, tough critters the farmers could turn loose in the woods to forage for themselves. They were called razorbacks because they were long and lean, with long legs and long snouts. People said that they "ran fast but grew slow." They had more bone and gristle than meat, and the meat

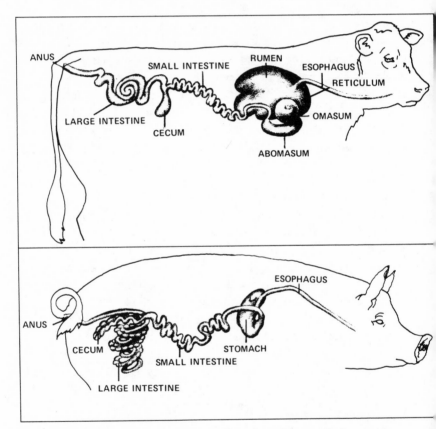

Figure 6–2. Digestive tracts of cattle and hogs. Their complex digestive systems enable cattle to convert the leaves, stalks, and stems of plants into meat and milk, but hogs, like people, can only use the fruits and seeds efficiently. Reproduced by courtesy of the Texas Agricultural Extension Service. The Texas A & M University System.

was all in the wrong places. Farmers in southwestern Ohio began importing better-quality animals for breeding as early as 1811, and they developed and stabilized a lard-type breed known as the "Miami Valley hog" that was similar to the modern Poland-China variety.

The area of small corn-hog farms in Ohio and eastern Indiana is one of the three traditional sections that persisted in the Corn Belt from the time it was first settled until after World War II. The eastern Corn Belt has had small farms that concentrated on hogs. The western Corn Belt, centered on eastern Iowa, has had medium-sized farms where cattle as well as hogs were fattened. The central Corn Belt in eastern Illinois has had large, cash-grain farms (Figure 6–3). This variety of traditional Corn Belt agriculture is reflected in the fact that five of every ten ears of corn are used to fatten hogs, three to fatten cattle, one is fed to other kinds of livestock, and one is the raw material for one of more than 600 industrial products.

The farming systems of the Corn Belt have changed dramatically since 1950. In Preble County, Ohio, for example, the acreages of the principal crops remained pretty much the same between 1924 and 1949, and the average size of farms increased only slightly (Table 6–1). Between 1949 and 1982, however, the average size of farms increased by 50 percent, and soybeans had largely replaced small grains and hay. Preble Country had shifted from a corn-hog farming area to a cash-grain farming area.

The rural landscapes of the Corn Belt have been slow to change, however, and they still reflect the traditional farming systems as well as the revolutionary changes. In Preble County, for instance, the large number of farmsteads indicates that farms were small. There are five to eight farmsteads per square mile, which would be consonant with farms of 120 to 80 acres. The buildings on many farmsteads are not well maintained, a clue that they are no longer used for farming and that the farmstead now serves primarily as a rural residence.

Most farmsteads in the upper Miami Valley have a single large barn and a variety of smaller outbuildings. The jumble of barn types reflects the diverse genealogy of the people who settled the area. New Englanders built simple

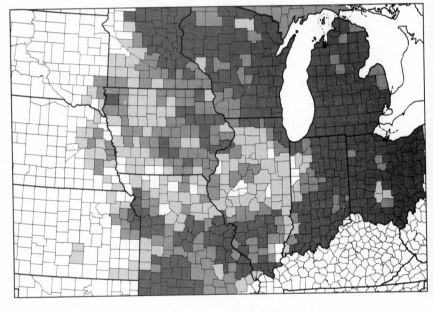

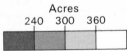

Acres
240   300   360

Figure 6–3. Average size of a farm in the Midwest in 1982. Farms are smallest in the eastern Corn Belt, in the Dairy Belt, and on the fringes of the Ozarks. Farms generally are slightly larger in Iowa, which was settled later, and they are largest in the Grand Prairie of Illinois, which required expensive drainage before it could be settled. The size of their farms influences farmers' decisions about the kinds of operations in which they can engage.

frame shells two stories high, with large double doors on either side leading to a central threshing floor. These barns originally had storage bays for grain, straw, and hay in either end, but they had no place for livestock. The animals were either housed in separate structures or they were left outdoors all winter. Later some farmers modified these barns by building stalls in one end beneath a hay mow, but the New England barn was designed for wheat farming. It was too simple for the corn-hog farming system.

People from Pennsylvania built the familiar, two-story forebay barns with ramps leading up to the threshing floor on the upper level. These barns were too elaborate for the corn-hog farming system, which had little need for the threshing floor, but did need abundant space for hay storage. The type of barn that was just right for the corn-hog farming system, neither too simple nor too elaborate, was the hay barn, with sheds on either side for livestock and machinery, that had been developed in the Great Valley of East Tennessee. It became the standard barn of the Corn Belt.

Over time the other types of barns were modified to make them more efficient and easier to use. I remember one perfectly good Pennsylvania German forebay barn in central Indiana that had been completely converted. The

	1924	1949	1982
Average size of farm (acres)	91	107	158
	Thousands of acres		
Corn	69	72	87
Wheat and oats	51	52	17
Hay	28	25	10
Soybeans	N.A.	2	50
Other crops	2	1	1

Table 6–1   Farm Size and Selected Crops, Preble County, Ohio

farmer had built a wall across the top of the ramp to block it off, and he had knocked out the upper part of one end and installed a hay fork. When I asked him why he had made these modifications, he looked at me with thinly concealed scorn and said, "Buddy, have you ever tried to turn around a team of horses and a hay wagon inside a barn?"

The most distinctive and the most common of the smaller outbuildings was the rectangular, wooden corn crib with slatted sides, which was a direct descendant of the single-pen, unchinked log corn cribs of the hills of Appalachia. It could be no more than 10 to 12 feet high in the days when the farmer had to heave ears of corn into it with a scoop shovel. The only way to make it larger was to make it longer. The introduction of mechanical elevators eventually enabled farmers to construct larger and more elaborate corn cribs, but they were essentially modifications of the simple single-pen crib of southwestern Ohio.

Fields in the upper Miami Valley were enclosed by stout fences of woven wire, or "hog wire," attached to steel posts and topped with a strand or two of barbed wire. Wire fences completely replaced the early wooden fences. It was easy to build zigzag fences of rails split from trees cleared from the land—a good man, it was said, could put up 200 yards in a day, and enclose 30 acres in a week—but they took up too much land, and the angles were prime breeding areas for pests and weeds.

In recent years many wire fences have been removed because they are no longer needed. The ideal layout for a Corn Belt farm, with a three-year rotation of crops that were used to fatten hogs, would have been three completely interchangeable fields. Each field would have been fenced securely to permit the farmer to move hogs into it with equanimity in the year when he wanted to hog off a corn crop or use it for pasture. In practice, because of the rectangular survey system, many farms consisted of four square fields of equal size, and farmers in level areas, such as the upper Miami Valley, grew two years of corn before the planted small grains and hay. Soybeans have replaced small grains and hay since World War II, and it is more efficient to fatten hogs in small pens than to turn them loose in the fields, so many wire fences are going the way of the split-rail fences that preceded them.

The upper Miami Valley is a level to gently sloping area, with occasional shallow stream valleys. The melting glaciers plastered it with thick sheets of debris that have weathered into deep fertile soils. Parts of the glacial plains are too flat. Extensive swampy areas had to be drained with buried tile drains and open drainage ditches before they could be cultivated. These areas were unhealthy in the early days, and they were used mainly for pasture. The early settlers disliked the level, poorly drained areas in the western part of Preble County and called them the "malaria flats."

The settlers used the trees that were growing on the land as ecological indicators. They knew that trees and other plants are closely related to the physical character of the sites on which they grow, and the existing forest cover told them what they might reasonably expect if they tried to grow cultivated plants on the same sites. They preferred "sugar-tree land," which had sugar maple and oak trees, because it was well drained, and they avoided "beech-tree land" because they knew it was swampy.

Most of the original hardwood forest has long since been cleared so the land could be placed in cultivation, but small rectangular patches still remain in farm woodlots that are in difficult or inaccessible areas, usually back from the highways. The total acreage of woodland is infinitesimal, but distant clumps of trees block out the horizon in every direction. In the old days every farm needed a woodlot for fuel, fencing, and building material, but the modern farmer buys such things at the lumberyard in town, and nowadays many woodlots look fairly moth-eaten. They have been used too long as convenience pastures, and grazing animals have eaten or trampled the saplings that would have replaced the older trees that are slowly dying off.

In 1983 I visited Gene Miley's farm, which is a mile southeast of Eaton, the county seat of Preble County. The land is tabletop flat, and the fields are fenced with woven wire on steel posts topped by a strand of barbed wire. The big, white, wooden forebay barn has "Miley" spelled out in green shingles on the gray-shingled roof. The forebay and the west end of the barn have been extended by skirtlike shed roofs to make more room for hogs on the ground level. Gene uses the upper level only for hay, machinery, and general storage.

The barn is on the other side of a gravel courtyard from the farmhouse (Figure 6–4). Next to the barn is a white, wooden hog building that also has storage space for 2,000 bushels of corn. Behind the hog building are four, cylindrical corn cribs of wire mesh that can hold 1,100 bushels apiece, and beside them is a long rectangular crib of wire mesh on telephone poles that can store 5,500 bushels of corn. The newest building on the farm is a gleaming white metal machine shed and workshop that Gene put up nine years ago on the eastern side of the courtyard. "One of these days we hope to build a new farrowing house where the sows can give birth to their pigs," he said, "because we sure do need one."

Gene is 47 years old. He does not believe in rushing into things, and he spends his money carefully. "The only money I've ever borrowed," he told me, "was when I bought this farm from my Dad. I paid it off last year, and this year I don't have any interest payments at all. I just

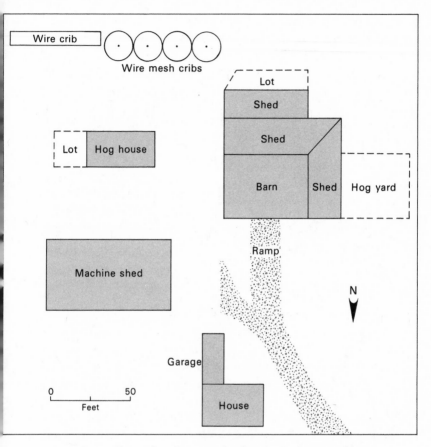

Figure 6–4. The Miley farmstead. The original Pennsylvania German barn has been modified almost beyond recognition by the addition of sheds to house more hogs. The new wire cribs suggest how much corn yields per acre have increased in recent years.

don't like to borrow money, and I don't see the point of spending a lot of money on new machinery and equipment. We're still using a 1963 truck to haul grain and hogs, a 1965 combine, and an eight-year-old corn picker. We have three tractors, one 72 and two 52 horsepower, a corn planter, a baler, a two-year-old stalk chopper, a grinder, and the usual wagons and cultivating equipment. None of it is real new, but it all works and gets the job done.

"I don't have much use for government programs, either, and I'm not participating in any of them. I didn't get into the Payment in Kind (PIK) program because the corn they would have given me wouldn't have been as good as the corn I can grow myself. Last year I ran short and had to buy some corn, and the hogs didn't like it nearly as well as what we'd grown here on the farm."

Gene is proud to be a hog farmer. "All I need is enough crops to feed my hogs," he said. He has 55 sows and 20 gilts (unbred females). He farrows 14 to 18 sows every month and a half. He only has 14 farrowing sheds, but figures some sows will come up barren. He raises 800 hogs a year, farrow-to-finish, and would like to get up to 1,100. His goal is to market them in six months at 220 pounds. A change of one cent a pound in the price of hogs will change his annual income by $1,760. He has a flock of 47 ewes, "just to keep the weeds down." He does not expect to make any money on sheep most years, but feels that he has to stay flexible and be ready to switch from one thing to another because costs and prices keep changing.

Gene Miley owns 120 acres and rents 110 acres on three other farms to grow enough feed for his hogs. His farm has 10 acres of wheat, 8 acres of alfalfa, 7 acres of woods that are pastured, and 95 acres of continuous corn. His normal corn yields run 120 to 150 bushels to the acre, "but 1983 has been a real bad year. We had a wet spring, and then it just quit raining, period. We had 15 straight days when the

temperature got over 100. We only got 75 to 80 bushels
from our best fields, a third to 40 percent less than our
normal yields, and we didn't even get our cribs filled."

All of the corn on his farm is picked and dried naturally,
not artificially, because the hogs eat naturally dried corn
better. He picks on the ear, and all of his storage is for ear
corn. He grinds his own corn and supplements it with purchased soybean meal. He grinds the corn cobs for bedding.
He has to shell the corn he grows on the rented land because it is grown on shares. On the rented land he rotates
corn, soybeans, and wheat. The beans and wheat are cash
crops. He plants beans in May, after he has planted corn,
and combines them in mid-September. He gets 35 to 40
bushels. He plants wheat as soon as the beans come off, and
combines it in July. He lets the field lie idle after wheat,
and then plows it up the next spring for corn.

"My family has lived within a mile of this farm for at
least 110 years. Glenn, my dad, was born in 1909 and raised
in a log cabin. He started farming on 40 acres, and rented
160 more. That farm was a mile east of here. He bought
this 120-acre farm in 1947. I bought him out when he decided to retire. I rent the biggest part of the land on this
road, but I don't like to get too far away because I don't
like to run my machinery and equipment on the highway.

"Carol, my wife, works in the office at the Producers
Livestock Association sales barn on auction days. She used
to work outside, and she did everything, worked the
ground, hauled corn, loaded hogs, but I haven't needed
her labor since Chris, my son, started working with me in
1975. He borrowed money to buy machinery." Chris Miley
was born in 1958. He and his wife Bertha have a son aged 3,
and another child is on the way. They live in Eaton, just a
mile up the road because they couldn't find a place to rent
in the country.

Chris hopes to sell his house in Eaton for part of the

down payment when he finds a farm he can buy. Last year he bid on a 36-acre farm. "I offered $2,500 an acre for it," he said, "which was more than it was worth, but it was right next door, and it had good buildings. It sold for $3,700. Farmland hasn't been selling much lately, though, and the going price in this area probably has dropped to around $1,500."

The Mileys farm 335 acres, which is more than twice the average size of farms in Preble County according to the 1982 Census of Agriculture. The census figure is deceptive, however, because the official definition of a farm is far too permissive, and the census includes many "farms" that are not genuine farms (Table 6–2). These "nonfarm farms" probably balance each other out when one compares census data for different counties, but they are quite misleading within any given county.

The Mileys have one of the smallest full-time farms in Preble County, and it is successful only because Gene is such a good and frugal manager. He uses a modified form of the traditional rotation on the rented land because his landlords request it, but he no longer rotates crops on his own farm because the old arguments for crop rotation have lost their persuasiveness. His land is so flat that he can grow continuous corn on it with no fear of soil erosion, and

	Number	Percentage
Total number of farms	1,319	100
Farms with less than 50 acres of land	450	34
Farms that sold less than $10,000 of farm products	566	43
Operators who worked off-farm 100 days or more	645	49
Operators whose principal occupation was not farming	636	48

Table 6–2 Farms in Preble County, Ohio, in 1982

he has at his disposal a whole arsenal of chemicals for maintaining soil fertility and for controlling weeds and insects.

The dependence of the Mileys on fat hogs as their primary source of farm income reflects the traditional importance of small corn-hog farms in Ohio and eastern Indiana. The farms in the eastern Corn Belt are so small that farmers have had enough time and labor to breed and fatten their own hogs, but most farms have been too small for successful beef cattle operations. Beef cattle have been more common on the medium-sized farms of the western Corn Belt in Iowa and adjacent states. These two mixed crop and livestock farming areas are separated by an area of large cash-grain farms on the Grand Prairie of east-central Illinois, where most crops are sold instead of being fed to livestock.

7

The Grand Prairie

The farms on the Grand Prairie of east-central Illinois have been so large that the farmers have had to concentrate their energies on growing crops, and they have had little time to spare for animals. They have sold their crops for cash instead of using them to fatten cattle and hogs. The Grand Prairie, as its name implies, was originally a vast expanse of monotonously level grassland. The first generation of settlers bypassed it, just as they bypassed other large prairie areas, and it was one of the last sections of the Corn Belt to be occupied.

The early settlers had good and ample reasons for bypassing the large prairie areas. These areas were treeless. They were inaccessible. They were poorly drained. They were plagued with mosquitoes and malaria. They had tough sod that was hard to break with the plows that were available. They were terrifying in the fall when the grass was dry because a chance spark or a bolt of lightning could set a fearsome, wind-whipped, prairie fire, a genuine wildfire that raced across the countryside for miles and destroyed all that stood before it.

The lack of trees was a problem, for settlers needed wood in abundance for buildings, for fences, and, most of all, for fuel. Wood for building and coal for fuel had to be shipped in to the prairie areas from great distances at considerable expense. The lumber and coal yard beside the railroad tracks in the small prairie town took the place of the woodlot on farms farther east.

Another reason the early settlers bypassed the large prairie areas was because they took the easy river routes into the Midwest. Indiana and Michigan are the only states in the midwest that do not take their names from rivers that were major routes of entry. Travel over land was far more difficult than travel by water before the railroad era, and the easy routes followed the stream valleys.

Few settlers ventured away from the river valleys and into the trackless prairie areas before railroads made them accessible. The railroads brought in people, lumber, fuel, farm machinery, and the more mundane necessities of life. Of even greater importance, the railroads provided a way of shipping out the goods that prairie farms produced. In the 1850s Congress authorized the donation of large acreages of public land to private companies to pay for railroad construction. The land-grant railroad company was given alternate square-mile (640 acre) sections of land on either side of its line. It sold this land to pay its construction costs. The railroad companies needed to attract settlers, both to buy the land and to produce goods that would generate traffic. They published lavish advertisements extolling the virtues of the prairies, and they actively recruited prospective settlers.

The settlers on the prairies had to develop a new technology for coping with the contrary environment. Their iron plows worked well in the forest soils farther east, but the heavy soils of the prairies stuck to the plowshares, and they did not "scour" properly. A prairie lad named John Deere made himself famous, and founded one of the nation's largest farm-machinery companies, when he began manufacturing "self-scouring" plows of steel that could be used to "bust" the tough grass sod of the prairies.

The prairies also required a new technology for making fences because they did not have enough trees for the familiar rail fences of the wooded areas to the east. The early settlers planted live hedgerows, even though hedgerows shaded out five or six rows of crops on either side. Good farmers grubbed out their hedgerows and replaced them after barbed wire had been invented in the 1870s, but even today a few hedgerows remain as mementos of the problems of early settlement.

Settlers in many prairie areas had to drain the land before they could cultivate it. The poor drainage of many prairie areas can be blamed on the selfsame glaciers that made the land so flat and fertile. The great ice sheets scraped away the topsoil and gouged out the bedrock as they ground slowly southward from Canada. This debris was frozen in the ice and dumped unceremoniously in uneven heaps wherever the ice happened to be when it melted. The irregular depressions between these heaps had no drainage outlets, and water collected in them. The normal processes of stream erosion had enough time to develop integrated drainage networks on the older glacial plains in the southern Midwest, but the younger glacial plains to the north were spotted with numerous lakes, ponds, swamps, marshes, bogs, sloughs, and mud puddles.

People drained the glacial plains of the Midwest by digging open ditches at the surface and by constructing mole or tile drains underground. They made mole drains by attaching a bullet-shaped, metal "mole" to the bottom of a narrow blade and dragging it through the sub-soil to create a small cylindrical tunnel at the desired depth. They made tile drains by digging a trench, placing drainage tiles end to end at the bottom, and covering them with earth. Drainage tiles are hollow cylinders of fired clay or concrete, one to two feet long and five inches or more in diameter. A tile drain is sturdier than a mole drain, but mole drains can be satisfactory in the right kind of soil.

Both types of underground drains carry water to open drainage ditches, which are the only obvious sign that the land has been drained. Many farmers, in fact, do not even know the location of their drains until one goes bad and has to be replaced. Drainage works are easy to ignore because their purpose is to make the land similar to adjacent areas, not distinctive from them, and the most compelling evidence of their success is the fact that you are not even aware of them.

Some of the most difficult drainage problems in the Midwest were posed by the plains formed on the bottoms of ice-dammed lakes so large that they were actually inland seas. These huge water bodies were created when the great mass of the glaciers blocked the natural northward outlets for the torrents of water that were pouring off the melting ice sheets.

When the ice dams finally melted and the waters were able to drain off to the north, the floors of the former ice-dammed lakes were exposed as astonishingly flat plains of extremely high fertility. Fine particles of silt and clay had settled to the bottoms of the lakes, and they provided the parent material for excellent soils, but the land was so flat that it required drainage on a grand scale before it could be cultivated. Once drained, the lacustrine plains are some of the most productive farming areas in the Midwest. The Grand Prairie is the best of the lot, and it may well be the finest farming area anywhere in the world.

The drainage of the Grand Prairie required far more capital than individual farmers could muster. Much of the area eventually was drained by entrepreneurs who needed to recoup their investments as quickly as possible. They sold their land in large blocks or divided it into large units that they rented to tenant farmers. The Grand Prairie traditionally has had the largest farms and the highest rates of farm tenancy in the Corn Belt. The land has been considered a good, safe investment because it is rich enough and the farms are large enough to provide a comfortable income for two families, tenant as well as land-lord. Tenancy actually is advantageous for the farmer, if he has reasonable security of tenure, because his capital is not tied up in land and he can use it productively. Farm tenancy on large holdings has become an accepted fact of life on the Grand Prairie: a tenant farmer on a large farm sees nothing strange about buying a smaller farm to rent to another tenant farmer.

The farms on the Grand Prairie have been so large that the farmers have had no time to spare for livestock, and they have concentrated on producing crops they could sell for cash. The best cash crop was corn, and the Grand Prairie has produced an exportable surplus of corn ever since the land was drained. Oats were a poor second in the cash-grain farming system. The traditional three-year crop rotation of the rest of the Corn Belt was never adopted in the Grand Prairie because the land was too valuable to use for hay. Oats could be fed to work animals before tractors replaced horses. They could be underseeded with a legume that was plowed under as a green-manure crop before farmers started using chemical fertilizers, but they were never very profitable.

The Grand Prairie was ripe to lead the way to the new two-year, cash-grain cropping system of corn and soybeans that has developed in the Corn Belt since World War II. In Ford County, Illinois, for example, corn has consistently occupied more than half of the harvested cropland, and soybeans have completely supplanted oats (Table 7–1). These three crops together have accounted for more than nine-tenths of all harvested cropland, and corn and beans alone accounted for 97.5 percent in 1982.

The railroads that served the Grand Prairie encouraged the development and persistence of cash-grain farming. In the early days they competed eagerly for the grain business by charging favorable freight rates for hauling corn.

The trackside area in every hamlet, village, and town on the Grand Prairie is dominated by huge grain elevators where the farmers bring their crops for sale. The electric sign at the grain elevator in Watseka, Illinois, even gives the current prices of corn and soybeans instead of the usual time and temperature.

The grain elevators are impressive partly because the countryside is so empty. The land is so flat that you can see across it for miles, and the skyline is broken only by an occasional relict hedgerow. The soil is deep, fertile, and almost inky black. Large fields of 80 acres or more are the norm, but they are still too small and too short: a farmer can cultivate a field at a rate of 54 acres an hour if it is laid out in rows a mile long, but he can only cultivate 42 acres an hour if it is laid out in half-mile rows because he has to spend so much time slowing down and turning around at the ends of the rows. Farmers have enlarged and elongated their fields in recent years. Few fields are fenced because few farmers have livestock, and most of the remaining fences are being removed.

The only trees are those that have been planted to shade the farmsteads. There are only two or three farmsteads per square mile, and they are small and simple. The landlord is reluctant to invest money in mere appearance, and the tenant is not willing to spend money to improve buildings he does not own. Every farmstead has a combination corn crib and granary for storing soybeans. Most have a

	1924	*1949*	*1982*
Corn	129	128	138
Oats	94	68	2
Soybeans	N.A.	27	123
Other crops	22	20	5

Table 7–1 Thousands of Acres of Selected Crops in Ford County, Illinois

machine shed and repair shop, but nearly all of the old horse barns have been torn down.

The traditional corn crib and granary of the Grand Prairie evolved from the simple corn crib of the eastern Corn Belt. Two rectangular corn cribs with slatted sides and straight shed roofs were built 10 to 20 feet apart, with the roofs sloping away from each other. The roofs were extended toward the center to form a single gable roof, and the upper part of the central portion was enclosed to form bins for storing loose grain. The ground level could be used to shelter machinery and equipment, including the elevator that the farmer used to fill the structure through a cupola on the roof. He hitched the power takeoff on one of his tractors to operate the elevator.

Russell Kirkham was farming three quarter-sections (480 acres) of rented land when I first visited his place in 1959. His farmstead was in northern Ford County, Illinois, three miles north of Piper City in the very heart of the Grand Prairie. The narrow, gravel, section-line roads were paralleled by drainage ditches 10 to 12 feet deep. I had to bump across a bridge over the ditch to enter his farmstead. The highest point on his farm, at the southern end, was six feet higher than the lowest point, a mile away at the northern end. A slope of six feet to the mile is one inch in 72 feet, flatter than the floor in your living room.

Russ Kirkham was strictly a cash-grain farmer and had no livestock at all. "In fact," he said, "the only thing on this farm with any blood in it is me and my wife." He grew 180 acres of corn, 180 acres of soybeans, and 120 acres of oats. He had no set rotation, but tried to avoid planting corn or beans on the same ground two years in a row. He grew oats only to rest the soil. He planted them with a legume seed in early April and combined them around the Fourth of July. The oats were mainly a nurse crop for the legume, which he plowed under in the fall as a green-manure crop.

He planted corn on the oats ground the next spring.

"A straight cash-grain farm can operate on only about three months' labor a year. There's a peak at planting time in May and June, and another at corn picking and plowing time in October and November, but there's not much work that has to be done in July and August. I usually manage to get away at least once a week for a round of golf. Of course there are always chores that have to be done, and in the summer a good farmer walks his fields to see how the crops are doing. If he finds any big weeds he pulls 'em up by hand."

Some of his neighbors said that Russ was a "B. C. F." farmer. "B." stood for beans, "C." stood for corn, and "F." stood for Florida. Around Christmas time, when he had finished his fall plowing, Russ and his wife usually took off for Florida for two or three months. "After all," he said, "there's nothing to do here on the farm until it's time to plant oats, and the winters in Florida sure beat those on the Grand Prairie.

"I do nearly all of the work on the farm myself. I have all six-row planting and cultivating equipment. Every row on the place is half a mile long, so I don't have to waste a lot of time turning around at the ends of the rows. The only time I really need help is at harvest when I'm operating the machinery and somebody else has to drive the tractor that hauls the wagons loaded with beans or corn back to the farmstead. My nephew, who's still in high school, comes out here and works for me after school."

Russ Kirkham told me that pretty good normal yields would be 80–85 bushels for corn, 30 for beans, and 40 for oats. He sells all of his corn, beans, and oats at the local grain elevator at the time of the year when the price is good. He never sells crops right after harvest because everyone else is selling and the price is lowest then. His corn crib holds 5,000 bushels of ear corn plus 3,200 bushels of grain, so he can store a good part of a normal crop.

Russ was a bit apologetic about the appearance of the farmstead. "My landlords are getting old," he said, "and they don't see any point in investing any money in the buildings. I can't say that I blame them, but I can't see the point of spending my own money to fix up buildings that I don't own. The landlords and I go fifty/fifty on the fertilizer, and we share the crop fifty/fifty. One of the biggest changes around here in the past ten years, apart from the shift to soybeans, has been a tremendous increase in the use of fertilizer on corn. The fertilizer really pays for itself in increased yields with the new hybrid seeds. We've always been big users of rock phosphate and limestone, but now we've started using commercial fertilizer, too. You could grow continuous corn on most of the land in this area if you fertilized it heavily enough.

"Land around here is selling for $400 to $600 an acre. If you've got good land, and you don't want to sell it, you'd better not put it on the market for less than $600, but no farmland is worth $500. It's way overpriced. It's selling for far too much ever to get it back farming. You'd really have a tough row to hoe if you had to live and pay the interest on a quarter-section."

A few years later, in 1964, I was passing through the area, and I dropped by to visit Russ. He told me he had cut back his operation to the half section he rented from the Yates' estate. I asked him why, and he said, "Well, about a year or so ago I got to complaining to the principal of the high school in Piper City about the way they were teaching American history. Finally he said to me, 'Russ, if you think you can do better, you've got yourself a job,' so I gave up a quarter-section of the land I had been farming, and now I'm the assistant principal of the high school and am teaching American history the way it ought to be taught."

Russ Kirkham, in addition to his other qualifications, has

a master's degree in education from the University of Illinois and a private pilot's license. His farm, like other cash-grain farms on the Grand Prairie, was highly specialized, completely mechanized, and making full use of new technologies as they became available. By 1982 most farms in other parts of the Corn Belt had adopted the model that was already fully operational on the Grand Prairie in 1959. They had become larger and more mechanized. They had switched from mixed-crop-and-livestock farming to specialized cash-grain farming.

The best place to start exploring changes in the Corn Belt, I decided, was the cash-grain farms on the Grand Prairie because in 1959 they had been far ahead of their time, harbingers of things to come in the rest of the region. In October, 1982, I went to look for Russ. The flat fields were crawling with self-propelled combines, some just finishing up the soybean harvest, others just getting started on corn. From the road they looked like giant, metallic insects rumbling along on huge front wheels and small rear wheels, with powerful mandibles chewing up the crops, glass-enclosed cabs for eyes, and unloading spouts jutting out at an ungainly angle on one side. I did not see a single corn-picker, but I saw plenty of old-fashioned, wooden corn cribs with slatted sides; they obviously were derelict, forlorn, and abandoned, replaced by squat new cylindrical grain bins of gun-metal gray corrugated steel. Oats and hay had vanished from the landscape, and not a fence was in sight.

Every five or ten miles I bumped across a set of railroad tracks. Railroad lines crisscross the Grand Prairie, and every few miles along each line there is a grain elevator, four, five, or even more massive concrete silos as tall as a four-story building. A few stores and a tavern huddle in the shadow of the elevator on the street that runs at right angles to the tracks.

I left the blacktop highway and turned onto the gravel road where Russ lived. At first I had trouble finding his farmstead. I drove up and down the road several times before I finally recognized his old corn crib. The shape was the same, but the end had been knocked out, sheets of rusting metal had been nailed over the sides, and it was being used to store small machines.

Everything else had changed. The old farmhouse had been replaced by a modern, two-story, olive-green house with brown shingles that would have been quite at home in any suburb. In back of the old corn crib stood a large cylindrical grain bin and two smaller bins, and to the left was a new, metal machine shed with blue-green sides and sliding white doors. Even the name was different. Nailed to the telephone pole beside the wooden bridge over the drainage ditch were signs advertising hybrid seed corn, and beneath them was a sign that said, "Wayne Morrison, Dealer."

I drove across the bridge and stopped a young woman driving a combine out to the fields. I asked her if this hadn't once been Russ Kirkham's place. It had, she said, but he had retired and moved into town, so I turned around and went looking for him there. His new home is an attractive, ranch-style house on a quiet street in Piper City.

Russ was as lean and lively as he had been a quarter of a century ago. He invited me in, and we sipped coffee while he brought me up to date. "I used to rent 320 acres from the Yates estate," he said. "In 1975 the matriarch died, and the heirs decided to sell the land. They made me a very attractive offer, but I didn't have the money. I'd been pretty busy teaching school, and my health hadn't been too good. I suffer from emphysema, and I finally had to stop smoking after 45 years. I wasn't interested, but my nephew, Wayne Morrison, wanted to buy it. He'd worked on the farm since he was in the eighth grade and was almost like a son to us. He could borrow enough to buy the north 160, and I was able to find a good buyer, a doctor, who was willing to buy the south 160 and let Wayne farm it on shares.

"I had to get out because there wasn't enough income to support two families, and I had Social Security plus my school pension. I like to go out and help Wayne on the farm, but the air out there seems to be heavier somehow than it is here in town, and I can't work as hard as I used to. My wife and I still go to Florida every winter. My mother is in the nursing home across the street, and I manage her affairs, which are farmland. I have to stay around until mid-January to take care of her taxes, and then we go to Florida for a couple of months. She owns 160 acres that she rents to Wayne. He and his wife do pretty well. They're keeping up their payments on a 20-year mortgage, and they have a seed corn business that brings them in a fair amount of money."

Wayne Morrison, 35, was adjusting the chopper he was going to use to shred the corn stalks left by the combine when I drove up, but he stopped working to talk to me. "That was my wife who was driving the combine," he said. "She helps a lot with the field work. We don't grow anything but corn and beans. Our normal corn yield is 150 to 160 bushels, and our bean yield is 45 to 50. Our break-even point is $2.55 a bushel on corn and $5.50 on beans, but you have to take what they give you for them. We can store 13,000 bushels of corn in the big bin and 2,500 bushels of beans in each of the two smaller bins. We take the rest to the grain elevator. The grain from the rented land all goes straight to the elevator. They weigh it and store it until the owner decides to sell it, which he can do with a phone call. He pays storage on it until he sells it.

"We used to have wire bins for ear corn, but they were

damaged by a big windstorm in 1967, and the insurance company helped us build the small bins. We put up the big one ourselves in 1976 when we started field shelling. They ran a pipeline across the back of the farm in 1967. The court ruled that the money they paid for the right of way had to stay on the farm, so that paid for the new machine shed. We built the house in 1978. My father-in-law and I did most of the work on it."

The Grand Prairie had changed less than I had expected. In 1959 this cash-grain farming area had been the bellwether of the Corn Belt, but by 1982 other areas had caught up with it. One of the most striking changes on the Grand Prairie had been a shift from farm tenancy to part ownership. Tenants had farmed 61 percent of the land in 1959, but in 1982 they only farmed 29 percent. Many tenant farmers, it seems, took advantage of the good prices of the early 1970s to acquire enough land for a permanent operating base. They rented the rest of the land they needed for their operations.

In 1959 farmsteads on the Grand Prairie had been a bit down at the heels. The area had not looked like the prosperous farming region it really was because tenants had been reluctant to invest in buildings on land they did not own. In 1982, however, the Grand Prairie fairly glistened, because the former tenants had adorned their newly acquired farmsteads with fine new houses, large new machine sheds, and gleaming batteries of new grain storage bins.

8

The Heart of
the Heartland

The westward sweep of American agriculture, which had germinated in southeastern Pennsylvania, attained its peak in Iowa, the heart of the nation's agricultural heartland. Early settlers bypassed the Grand Prairie to settle eastern Iowa first, even though it was farther west, so farms in eastern Iowa are smaller than the cash-grain farms of Illinois but larger than the small, corn-hog farms of Ohio and Indiana. Like the Illinois farmer, the farmer in Iowa had his hands full with crops during the summer, but like his brethren in Indiana, in the fall, when his crops were safely in the barn and in the crib, he could use them to fatten animals. With his larger farm, he could raise cattle. He could afford to pay a rancher to put bone and hide on lean feeder calves (6 months old) or yearlings (18 months old) that he would "finish" on corn with a bit of protein supplement. The farmer could use the corn that the cow's inefficient system could not digest to feed pigs and chickens.

After the Civil War the western Corn Belt was ideally positioned for these multiple roles. It lay between the ranching areas that were developing on the Great Plains to the west and the rapidly growing industrial cities of the East that needed more and more beef. At the end of the Civil War southern Texas was a veritable hive of beef cattle, which had been breeding cheerfully all during the war, but the area was a long way from any market. Cattlemen rounded up large herds of these cattle and drove them northward in great trail drives to cow towns on the railroads that were inching westward across the plains. By the time the cattle reached eastern slaughterhouses they were pretty tough critters, and their meat was not very palatable.

Some enterprising Texas cattlemen drifted on north past the railroads to the lush grasslands of the northern Great Plains. There they established ranches that could produce good, lean cattle, but these feeder cattle still had to be fattened on concentrated feed before slaughter. So feeder cattle from ranch country would stop off in Iowa to be "finished" en route to markets farther east.

The Iowa farmer fattened feeder cattle from the ranch country in a securely fenced feed lot. The farmer knew that cattle have less than perfect digestive systems, and some of the corn they eat passes straight through them. Hogs are not finicky eaters, and they eagerly devour the corn that the cattle fail to digest. The Iowa farmer knew that he could fatten hogs and feed laying hens free of charge, so to speak, behind each steer or heifer in his feed lot. The usual cow-sow-fowl ratio was one steer to two sows to ten hens.

Floyd Hoffman's farm was a fine example of an eastern Iowa corn-cattle-hog farm when I first visited him in 1958. He owned and farmed a quarter section of 160 acres, and he rented another 80 acres. He raised corn, oats, and hay. He fattened feeder cattle shipped in from a ranch in Wyoming, and he raised and fattened his own hogs.

Floyd's farm was on the "barren" (treeless) prairie ten miles east of Iowa City. The first white people who explored eastern Iowa found a plant cover of tall grasses, knee to waist high. The English language does not have a good name for treeless grasslands, and the early settlers called them "barrens." Later they borrowed the French word and called them "prairies." The prairies have enough rain to grow trees—four inches each month from April through September, with an annual total of 40 inches—and ecologists are still debating why they were not wooded. Many farmers have planted shade trees around their houses and windbreaks on the north and west sides of their farmsteads. The only natural trees are in the stream breaks along the rivers.

The silt-loam soils that developed under prairie grassland are ideal for growing corn. The decaying mat of grass roots adds enormous amounts of organic matter, or humus, to the soil, and gives it a dark brown color. The soils are derived from thick deposits of wind-blown dust, known as *loess,* that are three to nine feet deep. The loess, rich in plant nutrients, was deposited during the brief period after the last glacier had melted, but before the seeds of grasses and other plants could migrate in from unglaciated areas to the south to revegetate the bare ground. Powerful winds swept across the exposed surface, picked up the finer particles, and swirled them eastward in giant dust storms. The wind-blown dust accumulated like driven snow, and it rounded off a countryside that was already smooth. The broad, level uplands and long slopes are gentle enough to permit the use of farm machinery, but steep enough to be susceptible to erosion when they are cultivated with corn and other row crops.

Floyd's farmstead was on a low knoll on the west side of a gravel, section-line road. His substantial, two-story, white clapboard house was well insulated for adequate heating in winter and central air conditioning in summer. It was set back from the road on a pleasant lawn shaded by fine old oak and elm trees. The farm buildings behind the house were all painted red. They formed a veritable village of small structures, including a poultry house, a ma-

chine shed, an old washhouse, two thousand-bushel gra-
naries (one wood and one metal) and a corn crib with slatted sides.

The two largest buildings on the farmstead were the old
horse barn and the cattle barn. Floyd had used the horse
barn only for hay and general purpose storage ("That
means junk," said he, with a wink) since he switched from
workhorses to tractors. He used the cattle barn to shelter
cattle and hogs in bad weather. The cattle barn opened
directly onto the feedlot, which was enclosed by a sturdy
wooden fence of four unpainted planks nailed to massive
posts set firmly in the ground. The farm was divided into
seven 20-acre fields that were fenced with hog wire
topped by two strands of barbed wire on steel posts.

Floyd figured on a rotation of two years of corn, one year
of oats, and one year of a mixture of alfalfa and red clover
for pasture and hay. His normal corn yield was about 80
bushels an acre. He stored it in the crib as ear corn and fed
it all to cattle and hogs. He produced about 8,000 bushels a
year, and usually had to buy 3,000 to 5,000 bushels more.
His normal oats yield was about 50 bushels an acre. He
stored the oats in a bin in the old horse barn, where he
used to feed them. He seeded legumes under the oats. He
made hay until he had 4,000 to 5,000 bales, which was all
he could use.

"I sow oats about the first of April. I disc the corn stalks
twice before I plow the sod, and then I harrow after plowing to make a good seedbed. In my spare time I haul manure to the cornfields. I plant corn the first week in May.
When it comes up I spray it with 2-4-D for weeds, and I
cultivate it two or three times. I make my first cutting of
hay in late May, and somewhere along the line," he said,
with a twinkle in his eye, "I like to get in a little practice
with something called a fishing rod. The last week of July I
combine the oats, bale the straw, and make a second cut-

ting of hay. From the middle of October until the first of November we pick corn for two or three weeks, and then I can heave a sigh of relief and wait for Christmas. We don't have to work the way my dad used to because we can get things done faster. The mechanical corn picker has really taken the drudgery out of farming, and the picker-sheller and the silo are going to become a must on all farms."

Floyd Hoffman fattened 120 yearling cattle, 40 heifers, and 80 steers a year. They were bought for him by a local buyer who went west each summer and contacted the ranchers. The cattle arrived in the fall, and Floyd started them on stalks and the down corn in the field he had just harvested. After 45 days he put them on a full feed of 20 pounds of corn and a pound and a quarter of protein. The heifers fattened faster, they could be sold earlier, and so spread the risk. He sold them in February, and the steers from June until the first of October. He needed 75 to 80 bushels of corn to fatten each steer.

Floyd said that hogs were a must to follow feeding cattle because they ate the corn that the cattle wasted. He farrowed 20 sows in June and December, and normally fattened 300 to 400 head of hogs. He marketed 200-pound hogs at five and a half months. He told me that hogs were not as dependable a source of income as they used to be, and there were more diseases to worry about. Many feedlots had become infested because so many hogs had been raised on them.

"I keep 40 head of ewes mostly for weed mowing. A $20 lamb and a $5 fleece from a $3 cull ewe is a good bargain. A small flock of scavenger ewes is increasingly common in this part of the country. We have a farm flock of 200 pullets and up to 50 capons. We buy 250 chicks in the spring, and the pullets start producing at about five months. We sell about 60 dozen eggs a week for most of the year. We raise our own potatoes, which last 'til Christmas, but modern cellars are not fit for longer storage. We raise our table vegetables, and my wife cans jams, jellies, and tomato juice."

Floyd bought his farm in 1937. He paid $150 an acre, which was considered too high at that time, but he bought it so his daughter could go to a consolidated school instead of having to walk to a country school. His son Tom, who was born in 1932, had a year at Iowa State before he went into the service, and when he came back he wanted to go right into farming. Floyd helped him get started by furnishing the equipment, and Tom furnished the labor.

"Farmland around here is worth about $400 an acre. You couldn't buy land for that, but a fellow would be foolish to pay any more than that for it. Still, you've got to get bigger, and many farm wives have to work to make ends meet. An 80-acre farm used to be a one-man farm, but nowadays you couldn't live on one, and even a quarter-section won't support a family any more. Tom can make a living here, but he'll never be able to get ahead. In my opinion us little farmers are doomed."

A quarter-section did not seem like such a little farm to me, but Floyd was right. It turned out that 1958 was a good year to visit him because the farming system that had persisted and prospered in the Corn Belt for a century or more was about to be revolutionized. It started with better seed, such as hybrid corn. Farmers had to pour on chemical fertilizers to take full advantage of the genetic potential of hybrid corn, and they used chemical weed killers to reduce the need for cultivation. Their yields skyrocketed, but so did their expenses.

Farmers had to learn to specialize, to concentrate on doing what they could do most profitably. They switched from the traditional three-year rotation of corn, small grains, and hay to a new cash-grain rotation of corn and soybeans. They stopped fattening cattle and hogs, and they tore out the fences they no longer needed. They enlarged their fields for more efficient use of the larger and more powerful machines necessitated by the new cropping system. They replaced their picturesque old barns with angular, metal sheds large enough to house the new machines, and they replaced their old, wooden corn cribs with shiny, new, cylindrical grain bins of corrugated metal, often in batteries of two or more connected by a spiderlike network of metal tubes.

Modern farmers have had to learn to manage money as skillfully as they manage crops and animals. They keep track of their operations with computers, and they may spend more time sitting at their desks than sitting on their tractors. Brains have replaced brawn as the prerequisite for successful farming. The farm wife has been pressed into service to replace the hired man, and she must be able to do any job that has to be done on the farm.

The family farm has become a highly specialized business with a larger capital investment and a greater volume of sales than most of the stores on Main Street. The farmer has had to keep enlarging his volume of production in order to remain in business. Some farmers have expanded by buying more land, but most have enlarged their operations by renting land from their neighbors instead of buying it.

In 1958 Floyd Hoffman's farm had been a good example of a corn-cattle-hog farm in the western Corn Belt, and in 1982 I went back to find how it had changed. From a distance it looked just the same, but as I got closer the buildings seemed older, more tired, and more weatherbeaten than I remembered. The barns had once been painted red, but long exposure to sun, wind, rain, and snow had bleached them gray. Other buildings also needed paint, and they had not been maintained very well. The whole place looked a bit shabby and slightly the worse for wear.

A woman came to the door when I drove up to the house. I told her I was looking for Floyd Hoffman. "Five or six years ago he was real sick," she said, "and he had to have his leg amputated. He and his wife have moved into a trailer house in town. His son, Tom, that's my husband, has taken over the farm. He's gone to the grain elevator with a load of corn, but he should be back any time now."

I wandered around the place while I was waiting for Tom to get back. The feedlot, where hundreds of cattle and hogs had been fattened in their time, was deserted. Weeds boiled up luxuriantly from the ground those animals had fertilized so lavishly. Here and there a stalk of volunteer corn had sprouted from a kernel missed by steer and sow alike. Strong winds had torn loose much of the blue composition roofing nailed over the old wooden shingles of the feeder barn, and some shingles were missing. The rooflines seemed to sag, and the weathervane on top of the old horse barn tilted over at a crazy angle.

Tom finally came driving up on his tractor pulling an empty auger wagon. I told him that I had come back to find out what had happened since 1958 to a fine example of a traditional Corn Belt farm, and he said, "You came just in time. I've sold the place, and I'm leaving in March while I can still get out with my whole skin. I sold the farm, 160 acres, for $2,750 an acre. It would have been worth more as bare land, without a building on it, because the buildings will all have to be bulldozed down. I sold it to a neighbor who already owns 800 acres and rents 160 more. We've had some good years, but the grain embargo just about wrecked us as farmers. We were made the goats. This huge grain surplus is depressing prices. It costs me $2.69 a bushel to grow corn, and the price right now is $1.85.

"Ninety thousand dollars passed through this farm last year, but I didn't get to keep much of it. There's no way I can afford to buy the new machinery I need, and I can't afford to maintain the buildings. I bought this old tractor for $7,000, and now it would cost $30,000. Everything has gone up but the prices we get. We used to farrow 50 to 60 sows and sold 500 to 600 hogs a year, but they were all on dirt, and I had problems with lungworm. I couldn't afford to build the new house I needed to stay in hogs. I stopped cattle years ago because there wasn't any money in them. A small farmer with 60 to 80 head just can't compete with the big fellows. Since I got rid of cattle and hogs I've been farming the whole place on a straight corn-and-beans rotation. My corn yields average 140 bushels, and my soybeans average 40 bushels.

"Just think, sixty years ago anyone with more than 80 acres needed a hired man to help him work it, and now I can't make a living on 240. I've already bought a trailer house near Dad's, and we'll move into that in March. I'll take any job I can find."

Tom blamed his plight on government programs and low prices, but part of the problem might have been his own failure to grow when he might have grown, to expand his operation even if he had to stick his neck out and borrow big sums of money. Traditional Corn Belt farmers have a horror of indebtedness. At the time Tom's frugality might have seemed admirable, but in hindsight, which admittedly is always perfect, a loan at 6 percent interest when prices were good looks like it would have been a fairly smart investment.

In 1958, when Floyd Hoffman told me he thought small farmers were doomed, he had mentioned a neighbor, Anders Mather, who had kept expanding, and I decided to find how his farm had fared. Samuel Mather started it with 160 acres in 1851, and by 1939 Anders, his grandson, had built it up to 380 acres. Anders Mather was an outstanding cattleman. Each fall he bought 500 to 600 lean "feeder" cattle from ranchers in the West and fattened them to a market weight of 1,100 to 1,200 pounds. He had enough hogs to clean up the corn that the cattle wasted in the feedlot, but cattle were his first love. He grew crops because he needed them to fatten the cattle.

Initially he bought cattle through a local dealer, but in the mid-1940s he decided that he could get better animals if he bought them directly, so he traveled west to ranch country and bought what he needed. His neighbors were impressed by the quality of his animals, and they commissioned him to buy for them too. He became a major dealer in feeder cattle and bought them for farmers all over eastern Iowa. He traveled extensively in the ranch areas of Wyoming, Colorado, New Mexico, and Texas on buying trips.

His grandson, Kenny, was born in 1940. Kenny went to the University of Iowa for a year, but he hated it, and in 1960 he rented the home place from his grandfather. Kenny continued to rent it when his father bought it after his grandfather died in 1964, and in 1970 he bought his own 250-acre farm a mile to the east. He has bought another 170 acres since then, and rents an additional 220 acres, for a total of 1,020 acres.

In the fall of 1982 he told me that he did not plan to expand anymore. "I can't take any more debt," he said, "and I'm already pushing my labor supply. I'm full-time, and my dad and my three sons help part-time." A year later he was more optimistic. The price of land was about $1,000 an acre lower than it had been a year earlier, interest rates had dropped, and his son had started to get interested in joining the operation.

"I'm ready to buy if I can find land I like at a reasonable price," Kenny said. "I'll look around for land for sale, and if I find something I like I'll make an offer on it. Right now my farthest piece of land is six miles away. I would like to keep everything in a ten-mile circle. I would really like to keep it all in five miles, but I don't think I can manage that, because too much of the land is in strong hands. Farmers bought out nonfarm owners 10 to 15 years ago when prices were good, and today most farmers want to hang on to what they've got and buy more."

Kenny grows corn and beans, and sells them as cash crops. He has no livestock. "I have more land but need less labor than my grandfather," he said. "He would probably turn over in his grave if he knew there were no cattle on the place. I stopped cattle in 1970 because they weren't making any money. The development of irrigation and feed crops in the West killed cattle feeding in this area. I had a chance to buy this farm, so I liquidated the livestock operation to get a down payment. I stopped hogs in 1972 because I had no feed storage or facilities, and it would have cost too much to build what I needed. There are still lots of 300-acre corn-hog farmers around here. That old corn-hog combination can't be beat year in and year out as a way to market corn. For me the ship has sailed as far as livestock are concerned, but it's a good way to market seasonal labor, and I would advise my son to get into livestock if he wants to farm."

Data from the Census of Agriculture show how Cedar County has changed. In 1959 a fifth of the cropland in the county was used for oats, a fifth for hay, and a fifth for corn in the traditional three-year rotation; the other two-fifths was used for continuous corn (Figure 8–1). In 1982 a quarter of the cropland was used for soybeans, another quarter was used for corn in a two-year rotation with beans, and nearly half was used for continuous corn. By 1982 oats and hay had faded into the limbo that soybeans had occupied in 1959. Hogs provided a third of the county's total farm income both in 1959 and in 1982, but cattle shriveled from just over half in 1959 to just under a fifth in 1982 (Figure 8–2). In 1982 nearly half of the county's farm income was generated by sales of corn and soybeans.

Kenny Mather expects 150 to 160 bushels of corn and 50 to 55 of beans. His six-row combine can pick 1,000 bushels of corn an hour, with a bottleneck in hauling because he can pick corn faster than he can haul it in from the fields. He has storage capacity for 120,000 bushels of grain in four bins with a central grain drier.

"The hardest thing about grain farming is marketing," he said. "Your goal is to hit the top half of the market. I spend more time at the desk than I do on a tractor, maybe 10 to 15 hours a week. We have a CPA to handle our taxes. Jan does most of the bookwork, and our data are fed into a computer at Iowa State. They send you regular reports so you can compare your operation with others like it."

I asked him for some figures on production costs, and he said, "An acre of corn costs $100 to $110 for seed, fertilizer,

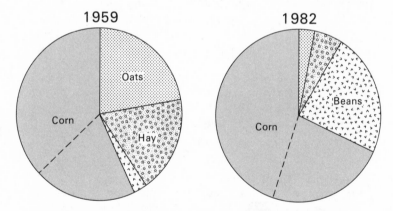

Figure 8–1. Principal crops in Cedar County, Iowa, in 1959 and in 1982. In 1959 about three-fifths of the cropland in the county was used for the traditional rotation of corn, oats, and hay, and about two-fifths was used for continuous corn. In 1982 about half of the cropland was used for a two-year rotation of corn and soybeans, and the other half was used for continuous corn. Reproduced by permission of *The Journal of Cultural Geography,* Bowling Green State University.

and chemicals, and $55 for equipment. The cost of land is
the major variable in determining the break-even cost of
crop production. It varies from zero if you own the land
and it is paid for, through $500 at 6 percent interest if you
bought it a while back, to $4,000 at 16 percent interest if
you bought it recently. Cash rent runs around $150 an
acre. The cost per acre is probably around $350. The
break-even price for corn probably is $2.75 to $2.80 a
bushel, and the current price is $1.85. Beans cost about $50
an acre less. The break-even price is around $6.00 a
bushel, and the current price is $4.85. You can't make
much money when you have to sell crops for less than it
costs you to grow them. All you can do is to tighten up your
belt, live on capital, and hope that prices will get better."

Kenny admitted that cash-grain farmers are underem-
ployed for six months of the year, but in winter he makes
management decisions on seed, chemicals, and fertilizer;

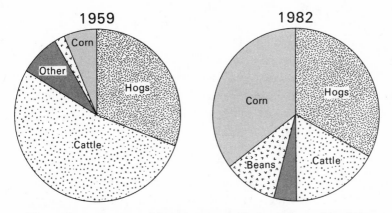

Figure 8–2. Sources of farm income in Cedar County, Iowa, in
1959 and in 1982. Hogs have retained their relative importance,
but cattle have been replaced by cash grain (corn and soybeans)
as the principal source of farm income. Reproduced by permission of *The Journal of Cultural Geography,* Bowling Green State
University.

attending meetings, seminars, and study groups; reads newsletters and farm magazines; and worries about tax management. He and Jan go to Colorado once or twice a year to ski, and they were thinking about attending a farm seminar in Innsbruck, Austria, in January 1984. They had heard that the skiing was good, and it didn't cost much more than the trip to Aspen.

Kenny told me he runs five to seven miles three days a week. I started to tease him about it, and he bristled, "Now wait a minute. Sitting on farm machinery, I don't get any more exercise than you do, and I need it. I feel better, I eat better, I sleep better, and I don't have any more trouble with stomach ulcers."

If any farmer ever has earned the right to have stomach ulcers it is Doug Magnus, who started farming in 1976. He could not have picked a worse time, because the wave of agricultural prosperity of the early 1970s had already passed its peak, and farm prices were in a long slide that was destined to last for more than a decade.

Doug's farm is on the gently rolling prairie of southwestern Minnesota. The countryside seems empty because farms and farm machines both have become so large. You can be sure that any clump of trees you see was planted to shade and shelter a farmstead, but like as not the buildings are gone because many farmsteads have been rendered surplus by farm enlargement. The average size of a farm in Murray County increased from 213 acres in 1950 to 323 acres in 1982. Even at the busiest seasons you rarely see machines in the fields because they do their job so quickly. They spend most of their time under shelter being prepared, maintained, or repaired. Few fields are fenced because fences are no longer necessary, and the highway department has encouraged farmers to remove them to reduce snow drifting during winter storms.

Doug Magnus was born in 1951. He went to South Dakota State University for two years, had two years in the army, including a tour of duty in Vietnam, farmed for a year, and then went back to SDSU. His father, Clarence, didn't think it made much sense for him to go back to college, so Doug had to finance his own education. His wife, Brenda, is a medical technician, but she took a job on a factory assembly line when he was in school. "I still couldn't have done it without the GI Bill," Doug said, "and it cost me a lot of money because those were good years for farming, but I'm glad I finished." He was such a good student that he was encouraged to go on to graduate school, but he wanted to get back to farming.

He started in 1976 by renting 400 acres. He rented another 160 in 1979, and in 1980 he bought a bare quarter, 160 acres with no buildings. He paid $2,100 an acre at 8 percent on a 15-year loan, but by 1982 the price of land in the area had dropped to $1,500 to $1,700 an acre. Doug rented his farm from his father, Clarence, whose grandfather was born in Sweden, migrated to Iowa, then moved to Minnesota in 1909 because he needed more land. Clarence started farming in 1950 on land he rented from his family, but he decided that there were too many Magnuses around him, and in 1962 he bought his farm near Slayton, nine miles west (Figure 8–3). He and Doug farm together, but they have no formal arrangement. They own a planter, a combine, and a sprayer together, but everything else is separate. They each own their own land, crops, hogs, and cattle. Clarence owns 440 acres and rents 300 more. Doug farms 540 acres, and together they custom farm 160 acres for an old fellow who hires them to farm it for him. They farm a total of 1,440 acres, and could not handle any more.

"The average cash rent around here is about $85 an acre," Doug said, "and not many are paying more than $100. I can remember the shock the first time somebody offered $100. I don't like cash rent because it has led to cutthroat bidding. Somebody will come in and offer more than you're paying when you have crops in the ground. Competition for land to rent is something fierce. I'm ashamed to admit I'm a farmer when I hear about the tricks some of them are trying to pull. Ninety-five percent of farmers are pretty honest, and I suppose that's the way it is anywhere. The bad ones are nice enough guys to talk to, but they can really pull some dirty tricks."

The Magnuses alternate corn and soybeans, and normally have 680 acres of beans and 680 acres of corn. Clarence grows 50 acres of alfalfa for silage and 30 acres of oats

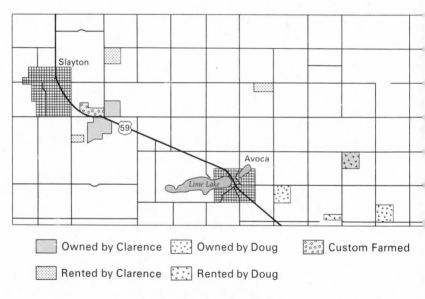

Owned by Clarence ⬚ Owned by Doug ⬚ Custom Farmed

Rented by Clarence ⬚ Rented by Doug

Section line roads are one mile apart.

Figure 8–3. The Magnus farms. Doug lives on 160 acres that he rents from his father, Clarence; he owns 160 acres, and he rents 240 acres more. Clarence owns 440 acres and rents 300 more. They custom-farm 160 acres for an elderly bachelor who is too old to farm the land himself.

on poorer ground to get straw for bedding. He cannot sell the grain for enough to pay the rent, not to mention the production costs, but he can harvest oats early and then put manure on the ground to make a good seedbed for corn. He would have no place to put the manure from the cattle feedlot if he didn't grow a little bit of oats.

They sell all the beans. They cut about 75 acres of corn for silage, dry and sell about 30,000 bushels, and feed the rest to cattle and hogs. They fill a bunker silo and a concrete silo with silage, and two concrete silos with high-moisture corn. Dry, shelled corn has a kernel moisture content of 15.5 percent, but high-moisture corn can be harvested when the kernel moisture content gets down to 35 percent. Dry corn is hard and flinty, but kernels of high-moisture corn are so soft you can punch a hole in one with your thumbnail. High-moisture corn has exactly the same feeding quality as dry shelled corn. They save the cost of drying it, but it spoils quickly, so they have to feed it as soon as possible after they take it out of the silo.

Doug feeds 75 cattle a year, and Clarence feeds 425. They buy cattle at a sales ring in South Dakota around the first of November, when the cattle are six months old, and feed them for ten and a half months. They buy at 500 pounds and sell at 1,150. "You lose money selling cattle from October to February," Doug said, "but you make money selling them from May 'til October."

Doug has pen space for 250 hogs. He buys feeder pigs because he does not have time to farrow them himself. He buys two-and-a-half-month-old pigs that weigh 50 pounds and fattens them in 120 days on shelled corn and commercial protein. He is constantly buying and selling hogs. He hauls a load of fat hogs to Sioux Falls, and brings back a truckload of feeder pigs. He does not keep his pens full all the time because he is too busy with crops in the spring and fall. Hogs are less risky than cattle, and he figures on making a $10 to $15 profit on each one. He bought his last bunch because he had a farm payment coming due. He contracted the selling price so he could be sure of making the payment. Hogs take the equivalent of about one day's work a week. Doug has depended on livestock to keep him productively occupied in winter, but the prices of livestock have not been good, so he would like to find something else.

Around the first of October he starts combining beans and normally gets 37 to 39 bushels. At the end of October he starts picking corn and expects 115 to 120 bushels. He would like to be done by the first of December, but it usually takes until Christmas. "We have to beat the deer to it," Doug said. "There are so many of them around here that we ought to have a second hunting season to get rid of them. You wouldn't believe the damage they can do to a field of corn.

"This spring," he said, "we spent a lot of time just standing by the tractors wanting to get into the fields, but they were too wet. We finally were able to plant corn, but a lot of fields were drowned out, and we had to replant them in beans. Then the crops were damaged by a severe hailstorm, and they were knocked down by high winds. In the summer we had bad drought conditions combined with high temperatures, and we got hit hard by corn borers, which eat through the stalks, make them fall over, and keep the ears from developing. Then the beans were hit by woolly bear caterpillars. I guess," he said, with a wry laugh, "it's been a pretty normal year."

In 1983 Doug and Clarence Magnus owned two six-row combines with interchangeable corn and soybean heads, which had a list price of $115,000 each; six tractors, with an average value of $18,000; three trucks, probably worth $32,000; a twelve-row planter worth $30,000; and enough plows, cultivators, wagons, augers, grain bins, feeders, and other equipment to add up to a grand total of $522,500. All of their machinery folds up so they can easily move it on the highway from one piece of land to another.

Doug and Clarence are partial to the brand of farm machinery that is painted green. The only piece of red equipment on Doug's place was an enormous plow with six huge moldboards. He seemed mildly apologetic about it and explained, "You'd hardly know the difference if I painted it green. The only reason I keep it is that Brenda likes it better than Dad's green six-bottom plow." I could hardly contain my astonishment. "Brenda!?" I exclaimed. She is lovely and lissome, crisply efficient but quiet and self-effacing. I could easily see her as a beauty queen or a model, but it was hard for me to envisage her tooling around the countryside in a tractor pulling that awesome hunk of machinery. "Do you mean to tell me," I expostulated, "that Brenda drives that monster?"

"Sure," said Doug. "She does most of the plowing. She also chops stalks and cuts silage. She drives the disc cultivator, but she refuses to drive the row cultivator. You really have to drive that straight between the rows, and she knows how mad I'd get if she took out twelve rows of corn or beans that would have to be replanted late in the season. Dad and I couldn't get along without her in the spring and fall because we're really stretched at planting and harvest time. My uncle also helps out at harvest. Dad and I run the combines, and they haul the grain in from the fields and dry it. She's one of the best truck drivers in the county."

The Magnus farm operation, like Kenny Mather's, is a modern Corn Belt family farm. Clarence and Doug farm 1,440 acres with additional family help at busy seasons. They are at the mercy of forces—weather, prices, interest rates—over which they have no control or even influence. "We probably have an investment here of $1.5 million,"

Doug said. "We gross $600,000 to $700,000 a year, but our interest payments alone run over $100,000. Last year mine were close to $50,000."

"We need to sell $300 an acre just to break even," he continued. "You can pick your expenses fairly close, but it's hard to figure what your income is going to be because you can't predict prices. I figure we need $15,000 a year to live on, $5,000 to pay the principal on my farm loan, and $20,000 to $25,000 for machinery.

"Things look pretty bleak right now for a young fellow just trying to get started farming, but remember that somebody's going to farm the land, even though they aren't making any money on it. They've just got to hope that sooner or later a good year is going to come along and pull them out. But farming's not for everyone. My brother-in-law wanted to farm real bad. He's not afraid of hard work, but he's not a good manager. He got into financial trouble a couple of times, and each time Dad and I had to pick him up and dust him off and get him started all over again. It cost us a lot of money both times. The third time he stumbled I told him to go back to college and complete the requirements for his teaching degree. He only had a few courses left to take. He likes to work with kids, and he's a good coach. Now he's teaching junior high, coaching the football team, doing a great job, and really enjoying himself."

The Magnus farm is near the western edge of the eastern United States, where rainfed agriculture slowly comes to a halt. Thus far the geography of American agriculture has been fairly straightforward and uncomplicated. An efficient system of mixed farming, based on growing crops in rotation and using them to fatten livestock, was transplanted to southeastern Pennsylvania from western Europe. Two streams of settlers carried this farming system with them when they worked their way through the

Appalachian uplands to the western waters. It took new root in southwestern Ohio, and thence it spread across the rolling glacial plains of the Middle West, where it flourished for a century and a half. It was economically successful and ecologically sound. It converted the Middle West into the Corn Belt, one of the human wonders of the world.

The majestic and apparently inexorable westward march of the Corn Belt farming system began to stutter and stall when settlers encountered a less congenial environment in Nebraska, Kansas, and the Dakotas. The soil was fertile, and it could produce good crops if it was adequately watered, but the rainfall was scanty and unreliable. An average year had no more than enough rain to grow a crop of corn, but few years were average. Many a year was so dry that the farmers could only watch helplessly as their crops withered away and the wind blew their precious topsoil eastward in great billowing dust storms. Farmers on the dryland margin eventually realized that wheat was the only crop they could grow successfully. They stopped trying to grow corn and hay, the other crops in the traditional rotation. Wheat became the principal crop of the Great Plains, by default, but even wheat failed in dry years. Some people think that wheat grows better in dry lands, but that notion is false. Wheat produces far better yields in more humid areas, but it cannot compete economically with other crops in such areas, and it is the only major crop that can be grown commercially on the margins of the dry lands.

Farmers in the wheat country of the Great Plains think and act like westerners, not like easterners. The vast treeless plains of wheat country are the eastern margin of the West, where the wide-open spaces begin. It is big country. Machines and farms both must be big because the semiarid land yields dividends so grudgingly in return for the expenditure of human effort. Wheat farms are larger than Corn Belt farms, and cattle ranches are larger than wheat farms. Farmers in the humid East measure their land in acres, but wheat farmers in subhumid areas think in terms of quarters (160 acres), and ranchers in dry areas talk about sections (640 acres, or one square mile). You know you have left the East when farmers start telling you about their land in quarters and sections.

9

Twice a Day
for the Rest
of Your Life

Environmental austerity limits the extension of the Corn Belt to the north as well as to the west. The principal environmental constraint to the north is lack of heat rather than lack of moisture; difficult glacial topography, sour infertile soils, and gloomy evergreen forests compound the problems of cool summers and a short growing season. On the plus side, some of the nation's largest cities are on the northern edge of the Corn Belt. The demand of their people for fresh milk has helped to create a Dairy Belt between the Corn Belt and the Great North Woods.

Corn, the basic American crop, likes long hot summers. The varieties of corn that were available before 1950 needed a growing season of 130 days or more to ripen into grain, and Minnesota, Wisconsin, and Michigan clearly were marginal areas for growing corn (Figure 9–1). In the northern parts of these states the summers were too short and too cool for any crop except hay. Even in the southern parts farmers had to fear that an early frost could damage their corn before the grain had a chance to ripen, so they harvested most of the crop as silage. In recent years plant breeders have developed new short-season varieties of corn, and cultivation of the crop has gradually edged northward, but the cool northern fringe of the Midwest still cannot compete with warmer areas farther south, and other crops, such as alfalfa, may be more profitable than corn.

Youthful glacial topography aggravates the disadvantages of the short, cool growing season. During the glacial epoch the moving ice scoured parts of the north down to the bare rock. It accumulated everything from large boulders to fine particles of soil as it oozed its way southward, and when it finally melted, it dumped this debris in confused tangles of hummocky knolls and ridges and poorly drained depressions. The debris left by the glacier is a heterogeneous mix, but in general it contains too many stones, too much sand, and not enough lime or plant nutri-

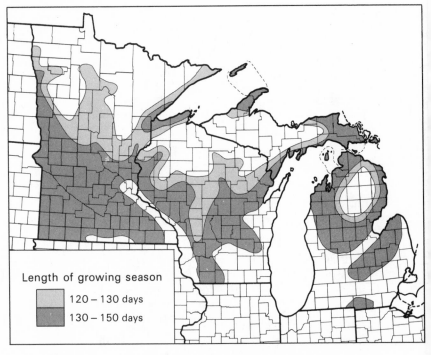

Figure 9–1. Frostfree period. The length of the growing season decreases northward across the Upper Lake states.

ents to produce fertile soil. Many slopes are so steep that they must be protected against soil erosion. Much farmland is wooded, with a high percentage of the cleared farmland used for hay crops or pasture (Figure 9–2).Lakes, ponds, sloughs, swamps, and peat bogs of all sizes and shapes own the depressions.

Splendid stands of centuries-old trees covered the jumbled glaciated areas in the northern parts of the Upper Lake states when European explorers first saw them (Figure 9–3). Lumbermen invaded this boreal forest in the nineteenth century. They felled the trees and then moved

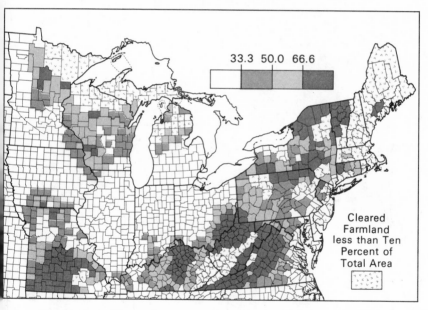

Figure 9–2.   Hay and pasture as a percentage of cleared farmland in 1982. Much of the farmland in the Dairy Belt, in the Appalachian Uplands, and in the Ozarks is so steep that it can only be used for hay or pasture crops. Many counties are so marginal agriculturally that less than ten percent of their area is cleared farmland.

on, leaving behind them a devastated, stump-covered, cutover wasteland. Today the forests of the Upper Lake states are mere spindly relics of their former grandeur. Bits of the primeval forest have been protected from lumbermen and forest fires in a few state parks and forest preserves, and they are truly awesome, but the cutover areas, for the most part, have degenerated into weedy jungles of birch and aspen, especially where they have been burned time and again because they have not been adequately protected against fire.

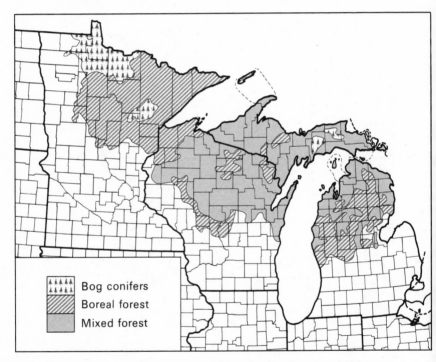

Figure 9–3. The Great North Woods. Northeastern Minnesota had a true boreal forest of needle-leaved, coniferous, evergreen trees, but these species were mixed with broad-leaved deciduous hardwoods in northern Wisconsin and northern Michigan.

The soils of the Great North Woods are shallow, stony, sour, and infertile. They are just as complex and vary just as much as the glacial deposits on which they have been formed. Sandy outwash plains that are excessively drained, droughty, and liable to serious wind erosion are next to poorly drained swamps and bogs where peat and muck accumulate because the lack of oxygen inhibits the normal microorganisms of decay. Much of the surface has only an inch or two of dark, sour, partially decomposed spruce and pine needles and other forest litter. Rainwater becomes acid when it percolates down through this surface litter, and it leaches the chemical plant nutrients from the soil beneath. The Great North Woods have few areas of truly good soil. Most of the land requires expensive drainage, liming, and fertilization before it can be cultivated with any hope of success.

The cards in the environmental deck truly are stacked against any attempt to farm at the northern edge of the Midwest: it has a short cool growing season, complex glacial topography, a cutover forest, infertile soils, and generally poor drainage. A farmer might be able to cope with any one of these environmental constraints if he could tackle it alone, but in combination they are just too discouraging. It simply does not make much sense, for example, to invest great sums of money in draining, liming, and fertilizing land if you can be reasonably sure that frost is going to wipe out your corn crop two or three years of every five. Small wonder, then, that only a tiny fraction of the land in the northern Midwest has ever been used for farming (Figure 9–4).

Although the Great North Woods may be poor for raising crops, one of the three major dairy-farming areas in the eastern United States is sandwiched in between the Great North Woods in Wisconsin and Minnesota and the corn and soybean fields of the Corn Belt (Figure 9–5). A second is in the productive limestone areas of southeastern Pennsylvania and Maryland near the great cities of the Eastern Seaboard. A third is in the valleys and on the flanks of the Appalachian Uplands of New York and Vermont. Dairy farming produces greater returns per acre in the first two areas, but farmers in New York and Vermont depend more heavily on sales of dairy products for their principal source of income.

The principal market for all three dairy-farming areas is the densely populated metropolitan and manufacturing belt of the northeastern United States. Fluid milk is bulky and perishable, and it must be produced close to those

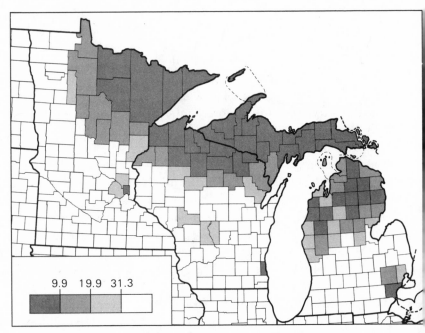

9.9   19.9   31.3

Figure 9–4. Cleared farmland as a percentage of total area in 1982. Environmental conditions severely constrain agriculture in the northern parts of the Upper Lake states.

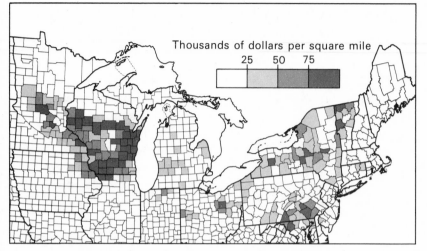

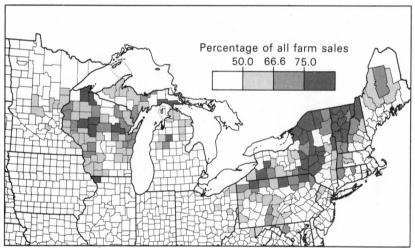

Figure 9–5. Sales of dairy products, 1982. The dairy-farming areas in Wisconsin, Minnesota, and Pennsylvania produce greater quantities of milk, but dairy farmers in New York and New England depend more heavily on milk as a primary source of income.

who are going to drink it. Dairy farmers who are too far from market can convert their milk into butter or cheese, which have a longer shelf life but do not fetch as good a price. Milk is seven-eighths water, and it is too expensive to carry any great distance. It turns sour very quickly, so it should be delivered to the customer within forty-eight hours of the time it leaves the cow. At one time milk was hauled to the cities each night by milk trains, which were notorious for their frequent stops. Today it moves by tank truck, but speed in getting it to the consumer remains imperative. Dairy areas have long been renowned for their good rural roads, which are essential for efficient milk collection.

Although the dairy areas are close to markets, the two northerly areas have to cope with such adverse environmental conditions as youthful glacial topography, infertile soils, and summers that are too short and too cool to produce grain crops competitively, but they are well suited to forage crops, such as alfalfa and corn for silage. Dairy farmers harvest much of their corn for silage rather than waiting for it to ripen into grain. They cut off the entire plant near the ground while it is still green, chop it into small pieces, and blow it into the towering cylindrical silos that are the hallmarks of dairy areas.

Alfalfa is a perennial crop that can be cut for hay two or three times a year. A good stand of alfalfa can be maintained for several years because the winter blanket of snow protects it from winterkill and frost heaving. Alfalfa yields the most tons per acre and the greatest nutrient value per ton, but it is merely the best of a long list of forage crops that includes many varieties of clovers, lespedezas, vetches, and grasses, such as timothy. A farmer can pasture cattle on forage crops during the growing season, but the animals waste part of the crop by soiling or trampling it, and farmers prefer to mow, store, and feed their storage crops.

Successful haymaking requires dry weather. The plants must be dried in the field for several days after they have been mowed, until their moisture content is less than 20 percent. A sudden shower can ruin a good crop of hay. Glass-lined, blue, metal silos, which have been developed since World War II, enable farmers to harvest and store forage crops when their moisture content is as high as 50 percent. The metal silos are gastight, and the plants in the silo remain in the same condition as they were when they were harvested. This moist product is called haylage. The new metal silos are expensive, but they permit the farmer to harvest his forage crops when they are at the ideal stage of maturity, and he does not have to wait for good haymaking weather.

Forage crops are the most economical source of nutrients for animals, but they are bulky in relation to their value, and they normally are fed to animals on the farms where they have been grown. Areas that have an abundance of forage crops usually have cattle of some kind because cattle and other ruminants have marvelously complex digestive systems that enable them to digest roughages. They can digest fibrous plants, such as grasses, and they can digest the fibrous parts of plants, such as their stalks, stems, and leaves. Poultry, pigs, people, and other animals are pretty much restricted to eating the seed and fruits of plants.

Cattle are common wherever environmental conditions inhibit crop production—where the slopes are too steep, where the rainfall is too scanty, or where the growing season is too short and too cool for crops to mature and produce fruit and seeds. One of the best ways to use steep slopes and subhumid areas is to grow grass for cattle to graze, and such areas are the domain of beef animals.

Dairy cows require better nutrition than beef cows if they are to produce copious quantities of milk, and the domain of dairy cattle is the cool moist areas where good forage crops can be produced. The best breeds were developed in the mild maritime areas of northwestern Europe, and dairy cows are happiest where the summers are not too hot.

A dairy cow is a far more complicated piece of business than many people seem to realize, and dairy farming requires much more than simply shoving hay and silage in at one end of the cow, pulling milk out of the other, and trying to dodge the inevitable by-product of the feeding operation. Cattle will cheerfully eat so many different things, and the kinds of forage vary so greatly from one area to another, that specialists in animal nutrition use the concept of "energy" as a yardstick for comparing the values of different feeding stuffs. They have developed some complex and highly sophisticated techniques for measuring energy, which is a bit like the way calories are used. Nobody has ever seen a calorie because it is not a specific substance, but everyone seems to be counting them like mad because they are a standard measure of the nutritional values the human body obtains from different foods.

Cattle can satisfy part of their daily energy requirements by eating roughages, such as pasture, hay, haylage, and silage, and they must eat roughages to keep their complex digestive systems in proper working order. A cow needs a basic maintenance ration of two to three pounds of hay (or its dry-matter equivalent) each day for every hundred pounds of body weight. At least a third of a cow's daily ration, by dry matter weight, must consist of roughages. Cattle can live on roughages, and a dairy cow can produce up to 70 percent of her milk potential if she is fed nothing else, but she needs concentrated feed if she is to fulfill her potential for milk production. She simply cannot consume enough roughage to produce milk at her maximum ability.

The dairy farmer must supplement the roughages in the cow's diet with more concentrated sources of energy, such as grains and other feeds that are rich in proteins, minerals, and vitamins. A dairy cow normally gets a pound of grain for each three pounds of milk she produces. The farmer begins by figuring how much of her energy and other needs she is getting from the roughages, and then he adds the most economical concentrate that will keep her producing milk with the greatest efficiency. The calculations get complicated so fast they make your head spin. For example, a pound of grain corn supplies the same energy as three pounds of alfalfa hay, alfalfa hay has nearly double the protein content of corn silage on a dry-matter basis, and a pound of silage is the dry-matter equivalent of only a third of a pound of hay. And so on. . . .

Fortunately, an individual dairy farmer does not have to make such calculations very often. He would know what roughages and what concentrates are available on his farm and in his area, and he would have worked out the best combination for his own operation. Corn silage, for example, is rich in digestible energy but low in protein; it meshes well in a dairy ration with alfalfa hay, which is low in energy but rich in protein. The dairy farmer knows from experience how much hay or silage his cows should get, and he does not need to weigh each cow to find out. He carefully records each cow's milk production, and he cuts back on her concentrates when her production starts to drop off because the feed she does not need to produce milk will wind up on her body as undesirable surplus flesh. (The beef farmer, on the other hand, gives his cattle as much grain and concentrates as they will eat in order to put flesh on them as rapidly as possible.)

Dairy cows, like other mammals, produce milk to feed their babies, and the dairy farmer must remove calves from their mothers as soon as they are born. He keeps the best female calves to raise as replacement heifers for worn-out older cows, and he sells the rest and all the male calves for veal. The farmer says a cow is "fresh" when her calf is born and her milk production is highest. Her milk production gradually declines, so after a standard lactation period of 305 days the farmer reduces her feed to "dry her off." She needs two months as a dry cow to regain her strength before she gives birth to her next calf. Normally she is worn out after four lactation periods, and the farmer sells her for hamburger. A fresh young replacement heifer takes her place in the milking herd. A farmer who is milking 40 cows will also have to feed 6 to 8 dry cows and 10 to 20 heifers.

A cow has a gestation period of 283 days, and usually she is bred again 60 to 90 days after she has had a calf. Farmers used to breed their cows to calve in late spring to take full advantage of lush summer pastures, but this led to a glut of milk in summer and a shortage in winter. The modern dairy farmer breeds enough cows each month to ensure a steady supply of milk throughout the year because he has learned how to store summer plant growth for winter feed. Two-thirds of the dairy cattle in the United States are bred by artificial insemination, which has improved the quality of dairy herds because frozen semen enables farmers to use better sires than they could afford to purchase in the past. The use of artificial insemination has also made dairy farms safer places because dairy bulls have a well-deserved reputation for being mean and dangerous.

The leading dairy-farming state in the United States is Wisconsin, which did not become a major dairy area until after 1880. Wisconsin was the third in a series of frontier areas where dairy farming replaced wheat farming during the nineteenth century as wheat production moved westward. Wheat was the principal cash crop and primary interest of the first settlers in the northern states. They kept a few scrub cows to produce milk for home consumption, but they expected the animals to fend for themselves. They turned the cattle out to graze in all seasons, and rarely sheltered them even in the worst winter weather. The natural grass pastures were good for only a few months in summer, and for the rest of the year the cows cost more than they were worth because the farmer did not have enough feed to keep them in milk.

The women of the farm preserved the surplus milk of summer by making it into butter and cheese. They made cheese by adding rennet—the membrane of a calf's stomach—to sour milk to make the curds coagulate. They pressed, molded, and cured the curds, and fed the watery whey to hogs. They skimmed the cream from whole milk to make butter and fed the skim milk to hogs. They put the cream in a churn and agitated it until the butter formed and could be worked out of the buttermilk with a wooden paddle. They salted the butter heavily to keep it fresh and packed it in wooden kegs. Some people fed the buttermilk to hogs, but others drank it themselves.

The farm wife took surplus butter and cheese, and anything else her family did not need, such as eggs, salt pork, lard, and smoked meats, to the country store, where she bartered them for groceries. The butter often spoiled, even though it was heavily salted, and people complained that it was unfit for human consumption and could be used only for grease.

The men of the family were preoccupied with growing wheat, which was the principal source of income in the early days. Harvesting and threshing wheat required long hours of hard work, and they limited the acreage a farmer could grow. When the grain was ripe he cut the stalks with a scythe, hauled them to the barn, spread them on the threshing floor, and beat them with a hand flail to separate the grain. He winnowed the lighter chaff from the heavier grain by tossing them in the air when a good breeze was blowing through the barn.

The first major commercial wheat-farming area west of the Appalachian Uplands was the Genesee country south of Rochester in upstate New York, which was settled largely by Yankees from Massachusetts and Connecticut. Flour from the Genesee country was one of the principal commodities carried eastward on the Erie Canal after it was opened in 1825.

The second major wheat-farming area was the "backbone counties" of northeastern Ohio between Canton and Mansfield. They were called the backbone counties because they were on the height of land dividing the streams that flow north to Lake Erie from those that flow south toward the Ohio River. The backbone counties were also settled mainly by Yankees and by Yorkers from upstate New York.

Cultivation of the same crop in the same ground year after year inevitably invited problems in the Genesee country, in the backbone counties, and later on, after it had become the newest wheat frontier, in Wisconsin. Farmers soon found that they were losing part of their wheat crops to Hessian fly, chinch bugs, midges, black stem, rust, smut, blight, and other insects and diseases. Continuous cultivation of wheat also depleted the fertility of the soil. Yields began to drop, but the price of wheat did not rise to compensate for the declining yields because bumper crops were pouring in from the newer areas to the west.

Some farmers sold or abandoned their farms and moved west to the newest frontier. Those who were determined to stick it out had to replace wheat monoculture with a more dependable farming system. They knew that rotating crops would help them to control the ravages of insects and diseases, and they adopted a modified version of the three-year rotation of corn, oats, and clover that had been developed in the Corn Belt. The growing season was so short that early frosts could ruin the corn crop, and they learned to harvest their corn for fodder instead of taking the risk of waiting for the grain to ripen. It would be too late to plant wheat after they had harvested the corn crop in the fall, so they would wait until spring and plant oats instead. They used clover as a soil-improving crop, and left it on rolling land for several years to protect the soil against erosion.

They cut hay from low-lying marsh and meadowlands that were too poorly drained to be cultivated, and they turned cattle out to graze on the slopes that were too steep. An acre of good pasture could provide as much feed as an acre of hay, and pasture was a big labor saver before modern machinery was available for making hay and silage.

Farmers in the northern states could not fatten cattle and hogs in competition with farmers farther south, who could wait until their corn was ripe and harvest it for grain, so they began to specialize in dairy farming. The construction of railroads after 1840 brought urban markets closer to the farm. Few milksheds extended out more than 50 to 100 miles from urban areas, but farmers farther away could produce butter and cheese. By 1850 upstate New York had become a major dairy-farming area, and the backbone counties in Ohio rejoiced in the nickname of "Cheesedom." The increased production of butter and cheese soon outgrew the capacity of the farmhouse kitchen, and groups of farmers began building cooperative processing plants that delivered a more reliable product.

Farmers no longer needed traditional barns with central floors for threshing wheat after they had switched from wheat to oats, and they realized that their dairy cows should have better shelter. Eventually someone got the idea of raising the entire structure, using the upper level to store hay and putting stalls for cattle on the ground floor beneath it to create a two-level barn. The two-level barn became the standard in dairy areas, with farmers from New York and Ohio carrying the idea with them when they moved west to Wisconsin.

Wisconsin enjoyed its own wheat bonanza during the early years of settlement. Many wheat farmers had kept a family cow or two, but they scorned milking as women's work. In 1837 *The Chicago Weekly American* derided Yankees as "a shrewd, selfish, enterprising, cow-milking set of men." By 1880, however, wheat was no longer a major crop in the state. Yankees, New Yorkers, and Germans, who were not ashamed of milking cows, turned Wisconsin into a dairy state after the wheat boom had ended.

A third of the state's residents in 1860 were natives of New York or Germany (Figure 9–6). The number of Yorkers declined steadily thereafter, but the influx of immigrants from Germany continued through 1890, and the state still has a strong German tradition. The New Yorkers introduced good dairy practices and bred better cows. They built cheese factories and developed contacts with eastern markets. The immigrants from Germany were prepared to adapt to the new country, and they were willing to do things that the natives found uncongenial: They were willing to accept the relentless drudgery of dairy farming. A dairy cow produces so much milk that the farmer must unload her every twelve hours, rain or shine, seven days a week. He must look forward to milking his cows twice a day for the rest of his life, with no thought of even a day off, much less a vacation. It is said that two things a dairy farmer will never need are a bed and a Sunday suit.

Many German immigrants were well-educated and highly skilled refugees from political oppression. They thought hard work was virtuous, unlike many American farmers, who "were not about to kill themselves." German women, unlike those of British background, were expected to work in the fields and to lend a hand with any chores that had to be done on the farm. The German immigrants were ready to work harder and to accept a lower standard of living than were those among whom they settled, while many American farmers refused to be "tied to a cow's tail."

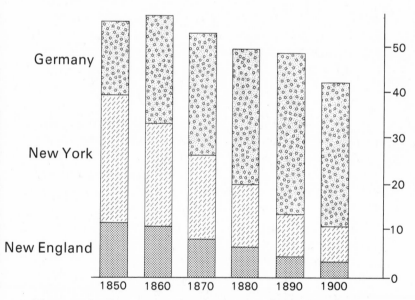

Figure 9–6. Percentage of non-native residents of Wisconsin between 1850 and 1900 who were born in Germany, New York, and New England. The state was first settled mainly by Yorkers and Yankees. It enjoyed a massive influx of immigrants from Germany after the Civil War.

Successful dairy farming requires a long-term commitment and a love of dairy cattle. The farmer must know each cow intimately, and he must be sensitive to her whims and fancies. The herd needs so much attention and affection that he cannot trust it to a tenant, so dairy areas have the lowest rates of farm tenancy in the United States.

Most dairy farmsteads are well maintained and attractive. The dairy farmer spends most of his time on his farm, and he likes to keep it looking nice. Dairy farms are smaller than grain and other livestock farms, and many are about the same size because there is a limit to the number of cows that a farmer and his family can milk. Small farms beget a dense rural farm population and prosperous market centers, which are closely spaced because dairy farmers cannot remain away from their cows for more than a few hours at a time. These market centers have cooperative creameries where butter and cheese are made. The tradition of cooperation to achieve common goals has carried over into the political arena, where dairy groups have been unusually effective.

The College of Agriculture at the University of Wisconsin played a prominent role in developing dairy farming in the state during the later part of the nineteenth century. Dairy farmers could not leave their herds long enough to come to Madison to learn about the latest scientific breakthroughs, so the university came to them. It created an extension service, with a county agent conveniently located in every county seat, to keep them informed.

The agricultural scientists at the university encouraged dairy farmers to switch from dry fodder corn to silage for winter feed. The farmers had been cutting their corn, stacking it in shocks in the field to dry, and feeding bundles of dry corn to their cows, which were bred to produce milk in summer. The scientists urged them to chop the green corn and store it as silage for winter feed, which would enable them to breed their cows to produce milk throughout the year. In 1900 Wisconsin had fewer than a thousand silos, but by 1914 nearly every dairy farm in the state had at least one and many farms had several. The first silos were built of wooden staves, but soon concrete became common, and cylindrical, concrete silos became the hallmarks of the Wisconsin dairy landscape.

Wisconsin dairy farms have continued to improve. Half of the state's corn crop was harvested for silage in 1924, but less than a quarter of it needed to be harvested for silage in 1982 because the yield of silage had increased from seven tons an acre to twelve. Today's farmer can grow enough corn to fill his silo on little more than half his acreage. When the silo is full he can wait to harvest the rest of his corn for grain because plant breeders have developed short-season varieties of corn that have greatly reduced the risk of losing the crop to an early frost. The availability of the new short-season corn varieties has enabled some dairy farmers to switch to cash-grain farming in the warmer and flatter parts of southeastern Wisconsin, where they can commute to city jobs and grow grain on the side, and on the flat prairie plains of southwestern Minnesota, where dairy farming has always been marginal.

Improvements in the technology of making and storing forage crops after World War II enabled dairy farmers to bring their pastures and hay crops to their cows, which is more efficient than turning the cows out to graze. Better forage harvesters enabled a farmer to "green-chop," or to cut just enough of his pasture or hay crop each day to feed his cows. He could keep the animals in a small enclosed feedlot near the barn and place their feed in a trough on one side of the feedlot.

Green-chopping had drawbacks. It could replace grazing, but it could not provide winter feed because the farmer could cut fodder only when plants were growing.

He had to cut feed every day, but he could cut no more each day than the cattle could eat. In rainy weather he got drenched: he ran the risk of getting bogged down in mud and of bringing in fodder too damp for feed. He had to start cutting fodder as early in the season as possible, when the plants were still immature, and he had to keep on cutting when they were overripe and past their prime.

Glass-lined, blue, metal silos combined the advantages of green-chop and silage by making the storing of haylage possible. The metal silos had push-button unloading systems, which automatically transferred haylage to the feeding troughs. A second metal silo enabled the farmer to store high-moisture corn. He did not have to wait for the grain to dry in the field and risk damage by an early frost. These metal silos had drawbacks: their expense and their visibility. The tax assessor could easily see them, he knew how much they cost, and he was sure to raise the farmer's taxes as soon as he saw one.

The improved, and more expensive, technology has forced dairy farms to become larger and more productive. The median size of milking herds in Wisconsin increased from about 20 cows in 1960 to about 40 in 1980. Better breeding, better feeding, and better management increased the average milk yield per cow from 6,000 pounds in a 305-day lactation period in 1940 to 8,000 pounds in 1960 and 12,300 pounds in 1980 for the same period. The best dairy cows produce more than 20,000 pounds of milk.

Wisconsin dairy farmers have consistently produced more milk than people could drink. A relatively small area near a city could produce all the milk its people could use, and milk produced in areas farther away was hauled to creameries and made into butter and cheese. In 1959 nearly every crossroads in Wisconsin seemed to have a creamery on one corner and a tavern across the road. Creameries in the more distant areas made butter because it is the most concentrated product and can bear the greatest transport costs. A pound of butter takes twice as much milk as a pound of cheese. Dairy farmers took skim milk and whey from the creamery back to their farms and used it to fatten hogs. Many also supplemented their income by raising cash crops, such as sweet corn, snap beans, green peas, or other vegetables that had little direct relationship to their dairy operations.

American tastes in dairy products have been changing. People worried about their waistlines began shifting from whole milk to lowfat milk or soft drinks in the early 1960s. Per capita consumption of milk in the United States dropped from 26 gallons in 1960 to 22 gallons in 1984, while consumption of soft drinks rose from 19 gallons to 40. For decades butter has been losing sales to cheaper margarine, which is made from vegetable oils such as cottonseed oil and soybean oil, but in the 1970s annual per capita consumption apparently leveled off at 4 pounds of butter and 11 pounds of margarine. In contrast, per capita consumption of cheese, especially mozzarella for pizzas, increased from 12 pounds in 1970 to 18 pounds in 1983.

Milk processors have concentrated their production in larger and more efficient plants, which can convert milk into the most marketable product, whether fluid milk, cheese, butter, or other dairy products. The old creameries at the country crossroads have been abandoned or converted to other uses.

Cows produce the most milk in the spring and early summer when they are grazing on the new green growth, but people drink the most milk in the fall and winter, when production is lowest. Urban milksheds used to expand in the fall and winter, when supplies were low and the price of milk was high, but in spring and summer, when dairy farmers produced lots of milk, they could hardly give it away. In the 1930s and 1940s the federal government developed a dairy program that was designed to alleviate seasonal fluctuations, to stabilize prices for farmers, and to ensure a reliable supply of milk for consumers. The price-support program works indirectly. The government does not pay farmers anything, but it buys enough surplus butter, cheese, and powdered milk from processors to enable the processors to pay farmers a guaranteed minimum price for their milk throughout the year. The national price of milk, the price guaranteed by this process, is based on the cost of producing it near Eau Claire, Wisconsin, in the heart of the nation's most efficient dairy area.

The government has established 45 fluid-milk market areas, known as "milk marketing orders," in which processors must pay farmers a minimum price for their milk. The price of milk in an order increases approximately 15 cents per hundred pounds for every hundred miles from Eau Claire to compensate for the transportation differential. Milk marketing orders have encouraged the development of inefficient dairy operations in the South and Southwest, and the major cost of the milk price-support program is paid by taxpayers to buy the mountains of surplus butter and cheese that are slowly molding in government warehouses.

The price the processor must pay the farmer is based on the price at Eau Claire, but it is actually set by Congress. Dairy farmers have a long tradition of cooperating for the common good. Perhaps it is not surprising that farmers who produce milk should know which side their bread is buttered on: the dairy lobby is well organized and efficient, and it is notorious for its lavish contributions to congressmen of both parties, whether or not they represent major milk-producing districts.

Congress steadily increased the support price of milk from $5 per hundred pounds in 1972 to more than $13 in

1980. The high level of price supports, plus the encouraging regional effect of milk-marketing orders, sparked a sharp increase in milk production after 1978. Between 1960 and 1978 milk production in the United States was nearly stable at 112 to 119 billion pounds a year, but it soared to 138 billion pounds in 1983, while consumption remained almost unchanged.

More than 10 percent of all the milk produced in the United States went straight from the farm to the processor to the government warehouse, and by 1983 the cost of the milk program had risen to $2.6 billion. The secretary of agriculture said it was "ridiculous." The director of the Office of Management and Budget said it was "scandalous." An editorial in the *New York Times* on December 3, 1983, identified "some basic tenets of official Washington: When milk or tobacco is involved, politics will usually overcome principle. The shrewdest political investors in the country may be the big dairy lobbyists, who have contributed more than $1.8 million in the past three years to Congressmen of both parties."

Eventually even Congress and the dairy lobbyists realized that milk production had to be reduced, but Congress moved with its usual deliberateness. It finally approved a voluntary reduction program that was so complex by the time it had escaped the toils of the federal bureaucracy that few farmers could understand it. Most simply ignored it. Dairy farmers know that milk production and price supports both are too high, but individual farmers are reluctant to sacrifice their own income by reducing production when their neighbors are not doing so, and lowering the support price will put marginal farmers out of business. Congress must resolve the problem it has created, but our elected representatives have a truly miserable record of protecting taxpayers against the sharply focused pressure of well-heeled lobbyists.

10

Back along
the Milky Way

Specialization has replaced diversification as the primary goal of good farm management in the latter half of the twentieth century. Once upon a time American farmers tried to avoid "having all of their eggs in one basket" by doing a little bit of everything, but modern farmers, to an ever-increasing degree, have had to concentrate on doing what they do best and on doing it even better. They have spun off their less profitable and less efficient operations to other farmers and to other areas.

Dairy farming neatly illustrates the increasing specialization and geographic concentration of American agriculture. In 1924 almost every farm in the northeastern quadrant of the United States had a family cow or two, but by 1982 it was cheaper for many farmers to buy their milk at the supermarket than to bother with a cow, and dairy cows were concentrated in a few favored areas (Figure 10–1). The milky way that once extended from Minnesota to Maine had coagulated into scattered clots in response to different environmental constraints and different economic opportunities. These differences are highlighted by dairy farms in Calumet County, Wisconsin; in Madison County, New York; and in Hampshire County, Massachusetts.

Figure 10–1. Dairy cows per 80 acres in 1924 and in 1982. In 1924 nearly every farm had a family cow, but by 1982 dairy cattle had become concentrated in a few favored areas.

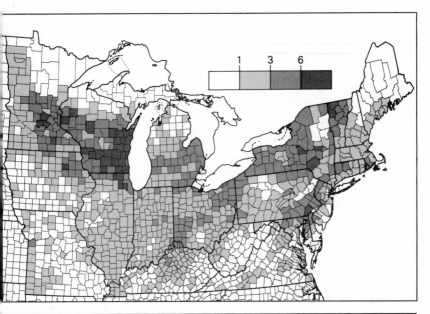

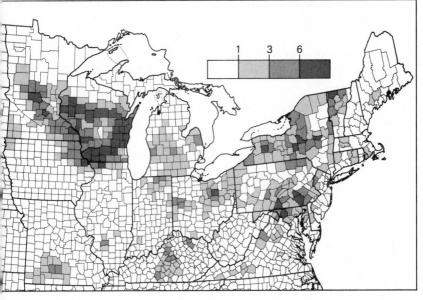

Calumet County is east of Lake Winnebago in the dairy country of northeastern Wisconsin. The rolling plains are the gift of the glaciers, which plastered the countryside with deep deposits of lime-rich debris when they melted. The soil is sweet and fertile, and most of the land is gentle enough for easy cultivation. Even the steeper slopes, where the snout of the glacier lay stalled for centuries, can be used for hay crops or for pastures, which protect the soil from serious erosion.

Fences have almost disappeared from the landscape, apart from a few forlorn relics, because it is more efficient to feed the cows at the farmstead than to turn them out to graze. The farmer who needs to enclose a field temporarily will "slap a hot wire around it." A single strand of electrified wire attached to white insulators on slender steel posts may not look too secure, but cows develop a healthy respect for it once they have satisfied their natural curiosity by licking one of the insulators with their damp tongues.

The farmsteads are handsome. Cylindrical, concrete silos tower over red barns with huge haylofts above solid masonry ground floors. Numerous windows admit light and air to the lower level where the cows are milked. Larger farms have several silos, including one or more gas-tight models of blue steel. Quite a few farms have their names emblazoned in bold white letters on the barn above the name of the owner. The substantial, two-story farmhouses are set on lawns that would be the envy of most suburban homeowners. Most farmsteads are graced by colorful flower beds, but few have kitchen gardens because it is cheaper to buy vegetables at the supermarket.

In 1982, when I visited Calumet County, Tim Meyer, the energetic county agent, told me, "The hard-working, German-Catholic farmers of this county have big families with lots of kids. Some of the kids drift away, but many of them stay on the farm. They really are tied to the farm. Going to town means Chilton [population 2,965] or Kiel [3,083], and the city means Fond du Lac [35,863]. When I was teaching vocational agriculture in high school, I took a bunch of them to a convention. Some of them had never stayed in a motel, and they didn't even know how to order a meal in a restaurant or pay for it."

Tim suggested that I visit Ralph Steiner. "Ralph is a good family farmer," he said. "He isn't hung up on appearances, and he doesn't feel compelled to spend money on new equipment—in fact, he makes a lot of his own. He's extremely innovative and is constantly trying out new ideas. He's a very smart farmer, and his boys were some of the brightest students I had when I was teaching high school."

Ralph's farm is on the gently rolling glacial plain four miles southwest of New Holstein. A bright plot of flowers adorns the lawn in front of the two-story, white, clapboard farmhouse. Next to the flower bed is a large, white sign with the oval red, white, and blue logo of the Lake-to-Lake dairy company; bold black letters at the bottom proudly proclaim "Ralph Steiner Family."

A gravel drive leads between the house and a one-story, red, wooden workshop on the right to the big, red barn, which is set end-on to the highway. An earthen ramp gives access to the upper level of the barn, whose lower level has been extended by a cinder block addition. A concrete silo and two blue, metal silos are near the barn, and a fourth silo stands in one of the securely fenced loafing yards. Each loafing yard has its own metal shed. Downslope is a concrete pond for holding liquid manure until it can be spread on the fields.

Ralph, 50, said, "I think my grandparents on both sides came here straight from Germany. I took over the farm from my dad when he was 65 and I was 25. He was born here. Alice and I have four boys and two girls. Bob wants to take over the farm. I think I may let him and start working for him instead of paying him to work for me. I was 25 when I started farming, and he should have the same chance I had. I don't want to make him wait until he's 40 before he starts on his own.

"I don't buy new machines; they cost too much. I love making my own machinery. I have made a tank spreader for liquid manure, a chopper wagon, tractor loaders, and an elevator. We do most of our own building and all of our own concrete work. We built the concrete manure pond. We made our own forms of plywood and two-by-fours and had to move them 75 times when we were pouring concrete. I plan to pave all of the loafing yards. I bought a concrete mixer that operates on the back of a tractor. It was a terrible headache until I installed a hydraulic motor, but ever since then it has worked like a charm."

The big, old, red barn has been greatly remodeled. The loft was used to store loose hay until 1948, when they started blowing dry, chopped hay into it, but barn fires from spontaneous combustion became so common that Ralph switched to bales. He no longer makes any hay, and he uses the loft to store baled straw for bedding. The ground floor of the barn has been expanded to twice its original size and now has 44 stanchions.

Ralph built the concrete silo 18 years ago to replace an old stave silo. It holds corn silage to feed the milking cows in winter. He built the second 13 years ago to hold silage for the heifers and dry cows that are in a separate loafing yard with its own shed. Three years ago he bought the 20-by-30-foot, blue, metal silo for high-moisture corn when he started producing more than he needed to fill the silos. The metal silos are easy to unload into feeding troughs in the loafing yard, the cows love the feed, and there is no waste.

Ralph Steiner owns 120 acres of land and rents 40 more from his brother. He had 80 acres of corn, 60 acres of alfalfa, 16 acres of oats, and was experimenting with 4 acres of soybeans. He grows corn for four years or until weeds get too bad and then seeds the field to alfalfa. At the end of September he cuts enough corn to fill the concrete silos. He gets 12 to 15 tons to the acre, and it usually takes about 40 acres to fill the silos. In mid-October he combines the rest of his corn when its moisture content gets down to around 25 percent and stores it in the small, blue, metal silo.

After corn he seeds alfalfa with oats or Canadian field peas. He had tried some direct seeding without a nurse crop, but it had to be done early because hot weather really hurts the young alfalfa plants. Planting peas is more efficient. The peas grow four or five feet tall, shade the alfalfa, and make good haylage. The cows like them better than oats, too. Ralph cuts his first crop of hay around the first of June, dries it in the field for a day or so, and then blows it into the big, metal silo. The second cutting takes place around the beginning of July. He gets about five tons of haylage in a good year. Some of the lower and wetter fields can only stay in alfalfa a couple of years, but the higher fields stay longer. Then he goes back to corn again.

The entire farm operation is based on the hay crop. You can maintain a cow on roughage, but she needs protein for growth, for milk production, and for the development of her unborn calf. Haylage is good roughage, and alfalfa has twice the protein content of corn. Corn silage also provides roughage, but you have to feed a cow additional protein when you are using it, and the high-producing cows need large amounts of protein. High-moisture corn is a good source of easily digestible energy to complement the protein in alfalfa. It is excellent for high-producing cows, but dry cows and heifers get too fat on it.

"I feed the cows their protein supplement when we are milking them," Ralph said. "We milk twice a day, at 6:30 A.M. and at 6:00 P.M., after supper. It takes about an hour and a half in the morning, but only an hour and a quarter in the evening because the time between milkings is shorter. I start milking the cows when they are two years old, and four 305-day lactations is pretty good. After that the cows go to make hamburger. I average about 16,000 pounds of milk per lactation.

"I milk more cows and buy more feed than I did when I started farming. I started with 24 cows, and now I'm milking 47. The way to expand is to add more cows and to buy more feed, not to get more land. You can buy feed and still make money milking cows. Many years it would be cheaper to buy corn than to grow it yourself, especially when you figure in the cost of machinery."

I went back to the farm after supper to watch the milking operation. Ralph has had cartilage operations on both knees, and Alice usually has to help him out with the milking, but this evening Barbara, 16, and Janet, 13, were on duty. They both knew exactly what they were doing. Janet opened the door to the loafing yard and the cows came jostling in. They found stanchions, where Barbara tied them up. Janet walked down the line with a big bucket of protein pellets under her arm. She put two cups of pellets in front of each cow. Barbara pushed a wagon of soapy water down the center aisle, and Janet washed each cow's udder. Barbara attached the milking cups to each cow's teats and coupled them to the overhead vacuum line. Janet came back with a wheelbarrow of high-moisture corn and placed two scoops in front of each cow, three in front of the high producers. Barbara removed the cups from each cow after she had been milked, gave her a slap on the rump, and sent her back out to the loafing yard.

Ralph kept moving cows from one side of the barn to the

other because only one side has a milking line. One old cow—"a problem cow," Ralph muttered—kept pawing at the milking cups with her hoof and trying to scrape them off. Ralph had to sit on an overturned bucket beside her while she was being milked to make sure she behaved herself.

Milking cows is an important source of farm income in upstate New York, just as it is in Wisconsin (Figure 10–1). Farmers in New York, in fact, depend on their milk checks for an even greater share of their income than farmers in Wisconsin, but they produce and sell less milk per square mile because nature has endowed their area less generously.

Upstate New York, like eastern Wisconsin, was remodeled by glaciers, but in New York the glaciers were bulldozing their way uphill into the uplands of Appalachia. They scraped smooth the hillsides and gouged out the valleys, some of which later filled with water to become the Finger Lakes. The ice, when it melted, plastered the countryside with stony debris, and it left a legacy of thin, sour, stony soils that require careful management.

Upstate New York is downwind from the Great Lakes, which cool and humidify the air masses that cross them. Gray overcast days are common, and precipitation is abundant, even excessive, but it is the kind of gentle steady rain that soaks the soil without eroding it seriously. Spring comes a week or two later in New York than it does in Wisconsin, the growing season is several weeks shorter, and the summers are a few degrees cooler. These differences may seem slight, as indeed they are, but near the margin even slight differences can be critical. It is possible to grow corn in upstate New York, but hay is a safer and more reliable crop. A quarter of the farmland is wooded, and two-thirds of the cleared land is used for hay and pasture.

Upstate New York was at the forefront of American agriculture between 1800 and 1850. Farmers made fortunes growing wheat and could afford the luxury of experimenting with new ideas. The area became a leading center of agricultural innovation, and prosperous farmers and small town merchants built fine mansions for themselves. The handsome small towns and pleasant rural areas of upstate New York have been discovered by affluent hordes from New York, Buffalo, Rochester, Albany, and other large cities within easy driving distance. Picturesque, old, country inns have been modernized, refurbished, and overpriced, and nearly every fine, old house in nearly every fine, old village seems to have been turned into a fine, old antique shoppe.

After 1850 the center of wheat farming shifted to the Midwest, and most farmers in upstate New York shifted to dairying. The standard dairy barn had room for only about 20 cows. The expansion of dairy herds in recent years has forced farmers to enlarge their old barns or to build new ones. The newest barns are one-story structures with loose housing for cattle. They are cheaper to build and cheaper to insure because hay fires are a constant threat in barns with lofts. A modern dairy farmstead will also have a shed for machinery and separate sheds and yards for heifers and dry cows.

Dave Timmerman has a splendid old, red barn on his farm on the steep hillside south of Oneida Lake in Madison County, New York. The barn is crowned by two large, louvered cupolas and flanked by two tall, concrete silos. Dave, 45, was too busy to visit with me, but he did take time to tell me that his great-great-grandfather had bought the 190-acre farm in 1855 and built the barn in 1875. Originally the barn had stanchions for milking 30 cows, but Dave has enlarged it to hold 60.

"Once this was one of the largest farms in the county,"

Dave told me, "but now it's one of the smallest. My grandfather had six hired men, but today it's just barely big
enough to support a single family, and I am going to have
to get bigger if my son, who is eleven, wants to join me
when he finishes school. This is no life for anyone who
doesn't want to farm, but I love it. When I am up there on
the hill driving a tractor, and I can look out and see Oneida
Lake glittering in the distance, I wouldn't trade places
with anyone in the entire world. I've got to be careful
where I'm looking, though, because some parts of the farm
are so steep that it's dangerous to drive the tractor downhill."

I had a surprise when I walked into the county agent's
office in Madison County in 1982 because I had never
before met a female county agent. As soon as I saw the
name of Kathryn E. Brown I realized that she had to be
better than good to be successful in a traditionally male
job. She was. She grew up on a dairy farm in western New
York, and she definitely knew her business.

"Our big advantage for dairy farming here in upstate
New York," she told me, "is growing good forage. We have
to buy concentrates, such as soybean meal, but we can
raise just about everything else we need to feed dairy
cows. You should have at least two and a half tillable acres
to feed each cow. We have good growing weather, with
mild, moist summers, but some years are a bit too moist.

"Three-quarters of our tillable ground is in alfalfa and
timothy grass hay. We cut the first crop of hay around the
end of May or as soon as it stops raining and get a ton to a
ton and a half per acre. Then we cut at six-week intervals
and get a ton and a half to two tons from the later cuttings.
We can bale the second cutting, but we chop the first and
third for haylage because the weather in early and late
summer usually is too wet to make good hay.

"After four years of hay we plow up the field and plant

corn. About a quarter of our tillable land grows ninety-day corn for silage. We get 15 to 18 tons to the acre and up to 25 in a good year. Some farmers combine part of their crop as high-moisture corn, but it's cheaper to buy shelled corn than it is to try to grow it here because our summers are fairly cool for corn. High-moisture corn also requires special storage facilities and good management because you have to feed it as soon as you take it out of storage. We grow corn on the same field for three years because we use atrazine to control weeds the first year, and the residue will kill alfalfa. Then we plant oats as a cover crop for hay, although some farms seem to grow oats simply out of tradition. We seed oats around April 15 and corn around the first of May.

"Pastures are not as important as they used to be, and most farmers use them mainly as exercise areas for their cows. A lot of our pastures have gotten badly infested with thornapples. The low bushy plants have pretty, white flowers that city people find attractive, but farmers detest them because they are a terrible pest. Cattle eat the fruit and spread the seeds with their droppings, and the plants are almost impossible to eradicate. Some thornapple thickets are so dense that not even a rabbit can get through them."

Kathe suggested that I talk to Harlan W. Jones about his operation. The Jones farmstead a few miles south of Cazenovia is as pretty as a picture postcard. All of the buildings except the two-story, white, frame farmhouse are painted brilliant red with neat white trim. They form a north-facing U around a large, tree-shaded lawn that could be used for a putting green. A brand-new machine shed and workshop on the west side faces the farmhouse and a range of low buildings on the east. The south end is closed by a T-shaped barn with windows on the ground floor and a huge gambrel roof. The barn is adorned with a lifesize painting of a black and white Holstein cow beneath "Jonesville" and above "H. W. Jones" in white letters. The stem of the T, which extends southward, is a single-story structure with many windows. It is flanked by two concrete silos and two blue, metal silos.

"I like bright red paint," Harlan said. "Regular barn paint starts to look three years old before it has even begun to dry. We built the barn in 1970 after the old barn burned. The old barn held 48 cows. The new one has 74 stanchions. I don't like free-stall housing because it's too dirty. My father said if you can't make a living milking 74 cows you'd better quit.

"He bought the 100-acre farm here for taxes in 1935, and he added 55 acres later. I bought the farm from him in 1964, and over the years I've added 215 more tillable acres. He's 84 years old. He was incapacitated by a stroke a year ago and lives in the former tenant house next door. I'm 47, and I have been farming since I was 5. My son is 20, and he isn't sure what he wants to do. My wife helps in the fields when I need her, and she keeps the farmstead spotless. I have a full-time hired man who lives in Cazenovia and a schoolboy who has worked for me part-time for seven years."

Harlan was milking 77 cows, with 30 yearlings and 42 heifers. His herd average is 18,115 pounds of milk. It was down a bit from a couple of years ago because he was milking two more cows. The average age of the cows was 49 months. He sells them for beef as soon as they start having breeding problems. If he can't make money on them, he said, nobody can, and it wouldn't be right to sell them to anyone else. He also sells bred heifers in the spring if he does not need them for replacements. In winter he keeps all of the cows in the barn where they are warmer. In summer those milking under 50 pounds a day are turned out to pasture, but he keeps the high milkers in the barn where it is cooler.

The milking parlor and the milk room are in the front part of the barn, beneath the hayloft, and the cows are in the one-story southern extension. They are tied to their stanchions with chains, and high-moisture corn and hay-lage are placed before each one. At milking time they are led out through the west yard into the milking parlor at the west end of the barn. Each cow gets five pounds of protein concentrate while she is being milked, and then she is led back to her stanchion and tied up again.

The milk room is in the east end of the main barn. Harlan took down from the wall of the milk room a certificate attesting that his farm is above the state average in milk yields, and he showed me a set of figures he penciled on the back. In 1965 he was milking 65 cows that produced an average of 15,900 pounds of milk with 575 pounds of butterfat, and in 1980 he was milking 74 cows that averaged 19,290 pounds with 703 pounds of butterfat.

I asked him to explain the phenomenal increase, and he said, "Good feeding, good breeding, and good management! My father used to say that a good bull is half a herd, but a poor bull is the whole herd because he can ruin you in no time at all. Your average drops for a little while when you start milking more cows, but you should soon get it back up to where it used to be. This really is a small farm, but I have to work 80 to 90 hours a week in summer and 40 to 50 in winter just to make a living. I have enough headaches with 77 cows that I don't want to get any bigger, but I may have to."

When Harlan said his farm was small he reminded me of Edward Montague's farm in the hills of central Massachusetts, which was truly small. Ed was milking 25 cows and just barely managing to hang on by his fingernails when I

first visited his farm in 1958. He owned 350 acres, but he
had to buy much of his feed because only 31 acres were
tillable. The rest of his farm was severely cutover hardwood forest or steep and stony pasture that, in his words,
was "hardly worth anything. Juniper has ruined all of our
pastures, and you can't kill it."

Ed's farm was in the hills west of the broad valley of the
Connecticut River. Roger Harrington, the ebullient
county agent in Hampshire County, said, "The hills and
the valley are like night and day. The hills are mostly
woodland and stony pasture, but parts of the valley are
pretty good farming country. Valley agriculture is mostly
specialty crops such as vegetables, potatoes, and tobacco.
The dominant stock in the agricultural section of the valley is Polish and Czech. The Yankees can't take the backbreaking work like the Poles, and they've been pushed
back up out of the valley and into the hills. The Yankee hill
areas are still pretty much dairy country, small farms that
are struggling to get by."

Ed's farm was five miles northwest of Northampton by a
narrow, winding country road lined with massive stone
walls. The handsome old two-story white frame farmhouse
sat next to the road. A large unpainted shed connected it
to a weatherbeaten old hay barn at the back. The barn had
been extended at right angles by a newer, but still old, red
dairy barn with numerous windows along both sides of its
ground floor. A 12-by-30-foot concrete silo stood in back of
the barn.

Ed, 56, told me, "My mother's people built this house
and the old barn in 1813. We installed plumbing in the
house in 1913. We cook with oil, but we still burn wood in
winter. We built the new barn in 1930. It has 24 stanchions,
but you need at least 35 to make a living.

"The cows get 40 pounds of corn silage, 15 to 20 pounds
of grain, and 3 pounds of citrus or sugar-beet pulp every

day, winter and summer. I buy citrus pulp from Florida or sugar-beet pulp from Minnesota in hundred-pound bags. The cows get 15 pounds of hay in winter, but only 5 to 10 pounds from May 'til October, when they're out on pasture. I'm paying $3 a hundredweight for grain, if you can imagine that! That's where all our money goes. That's why we can't pay our bills. What I'd like to do is to have the government throw out all these price supports so I could buy my grain cheaper.

"I have 6 acres of corn and 25 acres of mowing land. I rent another 30 acres of mowing land and draw hay from there for the winter here. At least that's what I put down when I have to fill out some government thing. Our mowing land is grass, mostly timothy. They get good crops of alfalfa down in the valley, but it winterkills too badly here in the hills. We put up about 2,000 bales of hay each year. Haying starts at the end of May. We make two cuts, and then let the cows eat the third.

"We put all of our manure on the corn land and fertilize it pretty heavily. We plant corn at the end of May and cultivate it three times to keep down weeds. When you cultivate this land you can't go twenty feet without hitting a big stone. It's sinful to walk across the field without carrying a stone in each hand. We have to clear them off the land every time we plow it. I have two tractors, but I cultivate the corn with horses. I keep the horses because I like them, and I need them for sugaring. We cut the corn for silage. I fill the silo on the rented farm, in addition to the one here.

"About all the woods are good for is maple syrup. We cut them hard during World War II. Before that we always cut a little, but we had to cut a lot when we had the hurricane because a lot blew over. It's got to grow a lot faster than it does now before it'll be any good. There won't be any

more cut as long as I live because it's not worth it. I tap
about 600 buckets of maple sap every March and get
around 100 gallons of syrup. People come right up to the
house and want to buy maple sugar when they see smoke
coming from the sugaring house, but most of it is already
ordered in advance for $5 a gallon."

"I do day work a good deal. I get a little road work every
now and then. I used to work in the woods full time and
did chores morning and evening, before and after work. I
still do a bit of hauling for a sawmill. Every Saturday night
in the fall I run hayrides for 40 people, mostly old Jewish
ladies from New York, who come up to a hotel in
Northampton when the leaves are changing color. I put
bales of hay in the wagon for them to sit on, hitch up the
horses, and keep them out for about two hours.

"A man would have to be crazy to try to farm here. We
don't make enough to pay our bills, only just enough
to keep the wolf from the door. I only do it because it's
the old home place. If I didn't love it I'd leave. If any of
my boys wanted to buy the farm, I'd sell it tomorrow for
$25,000."

In the summer of 1982 I went back to visit the Montague
farm. As I drove out from Northampton I was curious
about what I was going to find. Ed probably had passed
away, and I wondered what had happened to his farm.
Had another farmer taken it over? Or had the fields been
abandoned and allowed to grow up in brush? Or had the
land been sold to city people who wanted new country
homes on nicely wooded lots? Roger Harrington, the
county agent, had told me, "There's been an awful lot of
'back to the land' types up in the hills around here. They
buy five or ten acres, and then they come in here in their
combat boots and their bib overalls and their granny
dresses and they expect me to tell them how to make a

living on it." He snorted. "The only way they can make a living on it is to find regular checks from their parents in the mailbox."

When I finally turned the last bend I discovered that the farmstead had hardly changed. It had been spruced up a bit with new paint and bright beds of flowers, and the gravel barnyard had been blacktopped. The place looked better maintained than I remembered it, but obviously it was still a working farm. The sign on the mailbox said "Peter and Mary Montague." Now who, I thought to myself, is Peter Montague? None of Ed's sons was named Peter. My knock at the door was answered by a handsome young man with a lean face and a shy smile. "Oh," he said, "I'm Ed's grandson," and I immediately recalled the tow-headed nine year old who had been running around the farm when I was there in 1958.

"My grandmother died in 1965," he went on, "and my grandfather went all to pieces after her death. He sold off all of the cows, then a year later he bought some more, and then he sold those too. He leased out the tillable land, and sold some of the woodland to neighbors when he needed money to pay his bills. He got a job as a night watchman, and he hated it. He finally died in 1977.

"None of his sons was willing to take over the place after he died. My dad tried farming, then trucking. Now he has a job as a janitor, which he likes. I had always wanted the farm, but my grandfather refused to sell it to me because he said I couldn't make a living on it myself, much less raise a family on it. I worked on another farm for a while, and then in 1974 I bought into a partnership with John Serafin, who has the next farm down the road.

"I bought this place from the estate after my grandfather died. He used to say that he would be happy to sell it to any one of his boys for $25,000. Well, I had to pay $60,000 for it when the estate was settled, and I got it that cheap only because it was in the family. I lease the tillable land and the buildings to the partnership. There are just two of us, John and me, and we share fifty/fifty. Sooner or later one of us probably will buy out the other. I suspect it will be me because he's 57 and I'm only 33."

Eight years ago they were milking 20 cows averaging 12,000 pounds. In 1982 they were milking 42 averaging 17,600. The cows were all registered, which meant a better market for the surplus cattle. The milking cows were at John's farm because he had a bigger barn, and Peter kept 35 heifers and dry cows at his place.

Peter had 31 acres of tillable land, John had 44, and they rented 25 "in little bitty pieces scattered all over the place." They grew 40 acres of corn and 60 acres of hay. They cut all the corn for silage and stored it in a pit silo at the other farm. They cut 12,000 bales of hay each year. They made the first cut at the end of May, another six weeks later, and then they pastured or green-chopped it. Green-chop went straight from the field to the feed bunkers. They tested their silage and hay for protein a couple times a year and bought grain accordingly. A cow got a pound of grain for each three pounds of milk she produced. They no longer bought citrus pulp because the price got too high.

"The sugar house went to hell when my grandfather got old, and he sold off the equipment. I started again two years ago and have made about 100 gallons a year, but I've given most of it away. He liked horses, but I use tractors to haul the tanks of sap in from the woods. I still set buckets, but this year I tried some plastic tubes, and they seemed to work pretty well. I make a living by doing half a dozen different things. Our partnership owns a small sawmill, and we sell some lumber. We still burn wood in the farmhouse, and I sell a bit of firewood. I wouldn't think of buying it because it's too expensive, but it doesn't cost me anything when I cut it myself. I opened a gravel pit down along the stream, and I sell gravel to a local gravel company. I guess you could say that the dairy farm buys the groceries, and the gravel pays for maintenance and improvements. My wife, Mary, is a registered nurse, and she works part-time. That helps a lot."

After he had shown me around the place Peter told me he had to work on his lime spreader. Last fall he got a very good buy on some ammonium nitrate fertilizer. He had no place to store it, so he had it dumped on a piece of bare ground, and covered it with a sheet of black polyethylene. He does not own a fertilizer spreader, so he had used his lime spreader instead. The fertilizer had corroded the spreader so badly that it had seized up on him, and he had to struggle mightily to force it loose.

I admired his ingenuity and his skill at improvisation, whether at spreading fertilizer or at making a living. I was reminded of an earlier generation, and as I was taking leave of him I could not resist saying, "Peter, your grandfather would have been proud of you."

11

Out of the Back-
yard and into
the Fields

At one time most farmers and even many small-town people grew much of their own food. Nearly every backyard had a vegetable garden, a chicken coop, a shed for a milk cow, a grape arbor, and a few apple, peach, and pear trees. The family ate their fill of fresh fruits and vegetables in season, and they canned, dried, or preserved the rest for winter use. In an especially good year they might think about trying to sell some of the surplus, but a good year for one family was also a good year for everyone else. Often they could not even give the stuff away, much less find anyone who was willing to pay good money for it.

The backyard garden and orchard served the family adequately, but the growth of an urban population created a demand for vegetables and fruit of better quality. People are willing to eat almost anything if they have grown it themselves, but they insist on a quality product if they have to lay out their own money for it, and few backyard gardens and orchards were managed well enough to produce the uniform products of high quality that the city markets demanded.

Eventually some farmers realized that there was money to be made by specializing in the production of top-quality vegetables and fruits as field crops rather than as garden crops. Others abandoned their backyard gardens and orchards because they could buy better and cheaper vegeta-

bles and fruits at the grocery store than they could grow
themselves. The family garden now has become a hobby
that must be justified on grounds of recreation rather than
economy. The production of vegetables and fruits, even
more than dairy farming, has become increasingly specialized and concentrated geographically.

Nursery, vegetable, and orchard crops give high gross
returns per acre, and they require expensive intensive
care by the farmer, but our demand for them can be satisfied by far less land than the amount we need for the major
field crops, such as wheat, corn, and soybeans (Table 11–1).
American farmers can grow all the fruit we can consume
on an area smaller than the state of Massachusetts, and
they can grow all the vegetables we can eat on an area no
larger than Connecticut and Rhode Island. The nation's
limited acreage of nursery, vegetable, and orchard crops is
concentrated in three types of areas: areas that are near
cities, areas that have good natural growing conditions,
and areas that have specialized processing, shipping, and
marketing facilities.

Areas near cities are attractive places for intensive nur-

	Acres	Dollars ($000)	Dollars per acre
All crops	326,306,462	76,043,727	233
Nursery and greenhouse	466,231	3,821,196	9,196
Tobacco	931,655	2,545,141	2,732
Vegetables	3,330,637	4,145,446	1,245
Fruits and berries	4,886,534	5,924,585	1,212
Irish potatoes	1,268,213	1,509,830	1,191
Cotton	9,781,404	3,167,273	324
Corn	69,857,993	17,288,276	247
Soybeans	64,832,842	10,912,122	168
Wheat	70,910,293	8,053,318	114

Table 11–1 Gross Returns Per Acre from Specified Crops,
United States, 1982

sery and vegetable farms because they have such easy access to urban markets. Nursery farms in particular must be as close to their city customers as possible. The principal concentrations of nursery farms in the northeastern quadrant of the United States are near major cities such as Detroit, Cleveland, Cincinnati, Pittsburgh, and the great metropolitan belt called Megalopolis that stretches along the East Coast from Washington to Boston (Figure 11–1).

It is tempting to jump to the conclusion that areas near cities are cultivated intensively because they have unusually fertile soil, but the quality of their soil is rarely as important as the quality of their location. The area at the edge of the city is intensively cultivated, and the area of intensive cultivation keeps moving outward as the city expands, no matter what the quality of the soil. Each metropolis in the Northeast has concentric rings of specialized farms. The inner ring has nurseries and greenhouses, the outer ring has truck farms with roadside stands selling fresh vegetables and fruits, and the area beyond the outer ring has dairy farms. These rings keep expanding outward like a bow wave at the front of the expanding metropolis. The children of dairy farmers become truck farmers, their grandchildren develop greenhouses or nurseries, and their great-grandchildren sell their land to developers and retire to areas that have milder climates.

For example, the growth of New York City has been pushing an area of intensive cultivation steadily eastward on Long Island for more than a century. In 1860 10 percent of all vegetables sold in the entire United States were grown on the western end of the island, in Queens and Brooklyn. Today these boroughs are completely built up, and they have been for many years. The growth of the city has pushed the area of intensive cultivation steadily eastward through Nassau and Suffolk counties, and today farming is just barely hanging on at the very easternmost end of the island in Suffolk County. At one time Nassau and Suffolk counties were major potato-producing areas, but today Nassau County also is almost completely built up. Potato farmers in Suffolk County first switched to vegetables, ranging from asparagus to zucchini, which they sold directly to customers at roadside stands, but today the leading agricultural products of the few farms remaining in the county are nursery stock, flowers, and sod for suburban lawns.

Those who take pleasure in viewing with alarm have viewed with great alarm the loss of agricultural land to urban encroachment on Long Island, even though this loss is inevitable. Such people seem unaware that the farmland lost on Long Island can easily be replaced by the development of new areas of intensive cultivation in other directions from the city. For example, the deep peat deposits of former glacial lake beds have been drained and placed in cultivation in the Walkill valley 50 miles northwest of Times Square at the foot of the Catskill Mountains in Orange County, New York (Figure 11–2). The spongy muck soils are good for growing root crops, such as onions, carrots, radishes, and potatoes. They also produce fine crops of celery, lettuce, sweet corn, and other vegetables. In other parts of the urban periphery dairy farmers have switched over to growing vegetables, which they sell at their own roadside stands, while vegetable farmers have discovered that they can make more money producing and selling nursery stock.

Such changes are also taking place in New Jersey, which

Figure 11–1. Sales of nursery and greenhouse products in 1982. The principal concentrations of nursery farms and greenhouses were near major metropolitan centers. The farmers near such centers depended on sales of nursery and greenhouse products as a principal source of income.

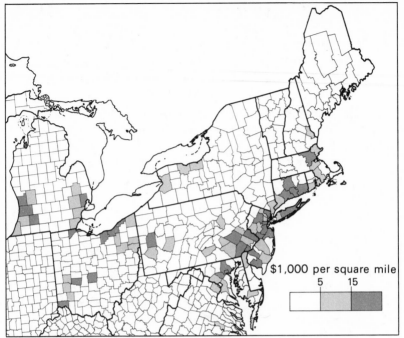

$1,000 per square mile

| | 5 | 15 |

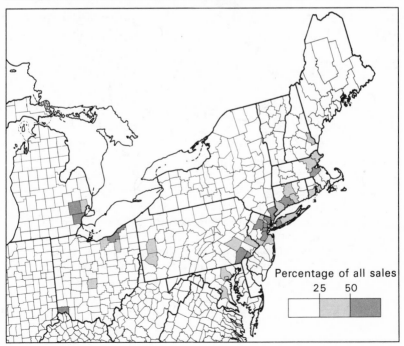

Percentage of all sales

| | 25 | 50 |

long has boasted that it is the Garden State. It might be described more accurately as the backyard of Megalopolis; some parts of the state truly are like gardens, but others seem more like places where people have dumped their junk. The state is part of a truck-farming belt that curves southward from Long Island to Chesapeake Bay. Farmers

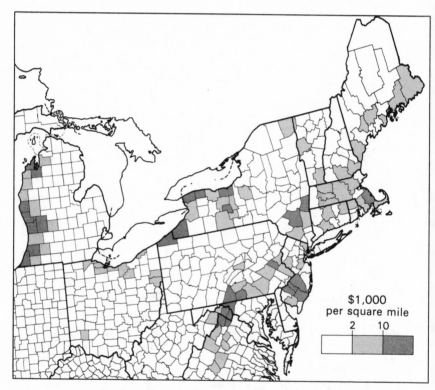

Figure 11-2. Sales of vegetables in 1982. Truck farms were a bit farther from major metropolitan centers than nursery farms and greenhouses. Truck farming was also a major activity on the sandy soils of the Coastal Plain in southern New Jersey and Delaware, on the Ontario Plain of upstate New York, on the rich muck soils of the Maumee Plain of northwestern Ohio, and in southwestern Michigan.

in this belt grow a variety of vegetables and berries for the great cities of Megalopolis. The sandy soils of the level Coastal Plain are well drained and easy to fertilize and irrigate. They warm up quickly in the spring, giving farmers a head start of two to three weeks over farmers farther inland.

Proximity to the urban markets of Megalopolis is an advantage, but the influx of new suburban residents creates problems for those who are trying to farm in the city's shadow. The new suburbanites can pay higher prices for land than farmers can afford, they force up taxes when they demand the level of public services to which they are accustomed, and they complain vociferously about such normal and necessary farm practices as spreading manure or spraying herbicides and pesticides. Vandalism has also been a problem near some new subdivisions.

Surprisingly enough, some truck farmers even have problems marketing their crops, despite their proximity to the markets of Megalopolis. Their farms are small, and they grow crops only during the warm months. They cannot compete with big growers in the West and South, who can produce larger and more uniform lots of better quality and who can grow crops for a greater part of the year. Supermarket chains prefer year-round contracts with large growers in the West and South instead of trying to deal with many small, local growers on a seasonal basis. Hundreds of small truck farmers on the fringes of Megalopolis have opened roadside stands where they can sell their produce.

Areas distant from large, urban markets have become important vegetable and fruit districts because they enjoy favorable natural growing conditions. Major vegetable districts in the northeastern United States include the Ontario Plain of upstate New York and the western side of the lower peninsula of Michigan, both of which are downwind from the Great Lakes, and the rich muck soils of the Maumee Plain in northwestern Ohio, southwest of Lake Erie (see Figure 11–2). Major fruit and berry areas in the Northeast include parts of Michigan and upstate New York that are downwind from the Great Lakes, the Hudson Valley of southern New York, the Cumberland Valley

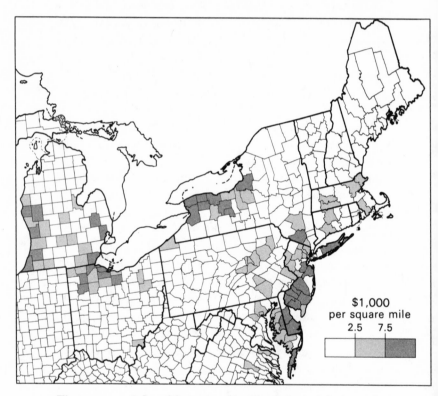

Figure 11–3. Sales of fruits in 1982. The principal fruit-producing districts were downwind from the Great Lakes in Michigan and upstate New York, in the Hudson River valley of New York and the Great Valley from Pennsylvania to northern Virginia, and in southern New Jersey. Many counties along the northeastern seaboard produced fruits and berries of lesser value.

of Pennsylvania and Maryland and the Shenandoah Valley
of Virginia, the cranberry bogs of the New Jersey Pine
Barrens, and the blueberry areas of southern New England. (Figure 11–3)

Areas east of the Great Lakes are especially favored for
fruit and vegetable production because they have relatively mild climates. Weather systems in the United States
move from west to east. The temperature of their air
masses is moderated when they cross the Great Lakes because water bodies warm up and cool off more slowly and
grudgingly than land areas. The Great Lakes warm up
cold air masses in winter, and they cool off hot air masses in
summer. The lakes warm up so slowly in spring and cool
off so slowly in fall that the seasons are retarded in downwind areas. Spring comes so late that fruit trees do not put
out tender buds until the danger of a sudden late frost is
past, and fall is so late that ripe fruit is rarely damaged by
an early frost.

The advantages of favorable natural growing conditions
in the fruit-growing areas have been reinforced by the
development of the specialized infrastructure necessary
for successful commercial fruit production: growers who
have learned how to produce quality fruit, a reliable supply of skilled labor, the specialized machines needed in
orchards and packing plants, processing facilities where
the fruit can be prepared for sale, salesmen and brokers
who have the necessary network of contacts with prospective purchasers, and adequate transport facilities for shipping the fruit to distant markets.

The requisites for a good fruit-producing district are
abundantly well satisfied in southwestern Michigan. The
area is close to Chicago and Milwaukee, it has the infrastructure necessary for producing and marketing fruit successfully, and it enjoys favorable growing conditions. The
beneficial climatic effect of Lake Michigan is reinforced by

the undulating topography of glacial moraines, which gives the orchards good air drainage. Cold air flows downhill, and the danger of frost is much less severe on slopes than in the low-lying areas where the cold air accumulates.

The early settlers around St. Joseph, Michigan, realized that the sandy soils of the moraines were better suited to fruit orchards than to field crops, and they quickly identified the lucrative market in the booming towns across the lake. They shipped the first boatload of peaches to Chicago from St. Joseph in 1840. The peaches arrived in much better shape than did peaches that had been brought by wagons on bumpy overland trails. Southwestern Michigan had captured the Chicago and Milwaukee market for peaches and apples by the time of the Civil War. The construction of railroads merely reinforced this dominance.

Six miles south of St. Joseph are the Nye Brothers' Orchards. Bold white letters on the north end of a large, red barn with a gambrel roof let you know that you are approaching "Nye's Orchards." The barn looms beyond a field of grapevines. An American flag snaps in the wind at the top of a flagpole in front of the house. The lawn is lined with bright flower beds. A high hedge and handsome ornamental plantings almost conceal the comfortable, story-and-a-half, red-brick bungalow.

Paved drives on both sides of the house lead back to the barn, which once housed dairy cattle. It has been transformed by the addition of cold-storage facilities and sheds. Behind it hulks a large, new, metal shed for open storage. South of the barn are a garage, a machine shed, and a well-equipped workshop. Nearby are a large cottage and five smaller cottages for summer workers. Everywhere, or so it seems, are high and tidy stacks of wooden pallets and slatted wooden boxes for fruit. The farmstead clearly is the heart of a working farm, but it has none of the clutter that seems almost inevitable around many farmsteads. Every-

thing is neat, in its place, and ready for inspection, almost
as if the farmstead were awaiting the arrival of visitors.

Gordon Nye, 38, and his brother, Dale, 34, run the farm
as a limited partnership, which they formed when their
father died unexpectedly in 1976. Their grandfather
bought the farm in 1915, and the Nyes have added convenient parcels when they came on the market. In 1982 they
had 520 acres of land in six different locations within two
miles. They grow 300 acres of soybeans and winter wheat
on land that is not suited to orchards, but their main prior-
ity is 85 acres of apples, 40 of peaches, 40 of grapes, 25 of
prunes, and 10 of pears. Gordon said that a family would
need about 125 acres of apples if they had nothing else to
live on and even that could be pretty tricky because the
price fluctuates so much.

The Nyes specialize in producing fresh market fruit.
They compete by emphasizing quality, and they get quality only by very good management, even if it means that
they have to go out on snowshoes in winter to prune the
trees. Apples are their principal crop. Peaches are more
reliable than apples because their price fluctuates less, but
they are harder to grow. They are more delicate, subject
to more diseases, and harder to harvest. They need a
higher, frost-free site, and they can be winterkilled if they
are too far from the lake. They prefer sandy soil, but they
will grow on clay if it is well drained. The Nyes have tile-
drained their land for peaches, but Gordon said the entire
farm only has about ten acres of really good peach land.

The Nyes grow grapes under contract with a processing
company and have them custom harvested with a special
machine. They harvest peaches from July twenty-fifth to
September twentieth, apples from mid-August to mid-
October, and pears for about three weeks starting around
August twenty-fifth. They get 700 to 850 bushels an acre
no matter what the fruit. Two men work for them full-

time, two older men work part-time, they get local help
when they need it, and they have a steady summer harvest
crew of fourteen Spanish-speaking workers from Texas.

The Nyes sell about a third of their fruit to packinghouses that prepare it for sale to supermarkets, but they
specialize in direct sale. Much of their business is pick-
your-own. They started to advertise in Chicago newspa-
pers in 1973 and have built up a mailing list of regular
customers to whom they send a letter in early summer and
another around Thanksgiving because they sell special
Christmas boxes. "This is a tremendous location for
U-Pick," Gordon said, "because our customers are mainly
from Chicago, which is only an hour and a half away, a nice
day's outing for the family. We have picnic tables in the
orchard where they can have lunch." One of their neighbors even sells trees to people. He takes care of the tree,
and tells the owners when their fruit is ripe and ready to
be picked. He lends them a table and they can picnic
under their very own tree.

The Nyes are open for U-Pick five days a week, 9 A.M.
until sunset in August and September. People check in at
one booth and pay by the bushel when they check out at
another. Most of the business is on weekends. The Nyes
have two U-Pick orchards, but open only one at a time, so
they can work in the other. They also have a roadside market, Nye's Apple Barn, two miles north of the farm at Exit
27 on Interstate 94 (Figure 11–4). It is open seven days a
week from June through September. They use surplus
labor at slack times to grow about 20 acres of vegetables—

Figure 11–4.   The Nye farm. The farm market is at Exit 27 on
Interstate 94, which is only an hour and a half from the Chicago
Loop. The farm, two miles to the south, has a wondrous variety
of apples for those who choose to pick their own and a picnic area
where they can enjoy their lunch.

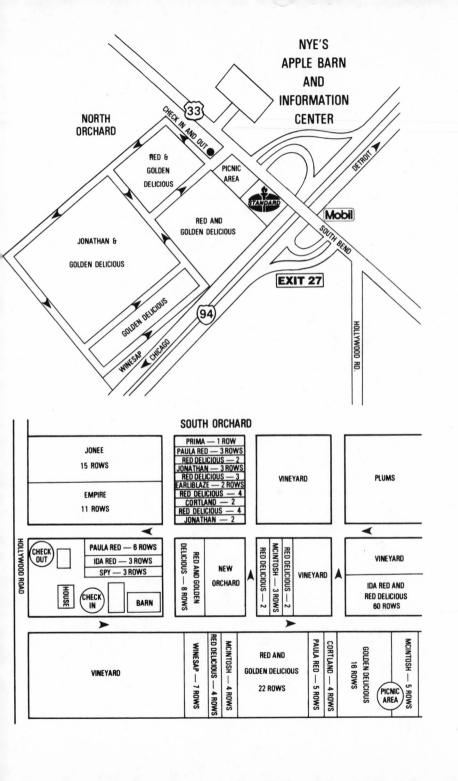

sweet corn, tomatoes, squash, pumpkins, watermelons, and cucumbers—for sale at their stand.

"St. Joseph is starting to squeeze in on us at the northern orchard near the stand," Gordon said. "We've divided part of the property into fourteen lots and are trying to sell them, but right now Michigan is pretty depressed, and they aren't selling very fast. A realtor is handling the development for us because we don't know enough about it. Sometimes we have trouble when we're moving equipment or spraying, but in a way it's nice to have people living near the orchards. They keep an eye on them for us, and they let us know if they see any suspicious strangers prowling around in them."

Urban neighbors can also be a problem for truck farmers, as they are on the flat, sandy lands of the Eastern Shore, which lies between Chesapeake Bay and Delaware Bay. Alan Wilber, who farms just north of Salisbury, Maryland, told me, "I have to be careful when I'm spraying herbicides because people are afraid I'll get some on their lawns and kill their grass. Once some people complained that my spray had killed their rose bushes after I had been working in a field near their house, and all I had done was plow. In fact, I hadn't used a single drop of spray in that field for at least six months."

I first visited the Wilber farm in 1959, the year before Alan was born. His father, Don, then 42, was raising truck crops, grain crops (corn and soybeans), and broilers. "About a third of my income is from truck crops, a third is from grain, and a third is from chickens," he told me. "That's a pretty good mix because truck crops are always a gamble. The yields are fairly satisfactory in most years, but the prices fluctuate something awful."

Don had 13 acres of watermelons, 6 acres of cucumbers, and 2 acres of cantaloupes. He had formerly grown tomatoes and sweet potatoes, but tomatoes and watermelons came to harvest at about the same time, and the light sandy soil was better suited to watermelons. The land is so flat that slope wash is not a problem, but the soil is so sandy that farmers do have to be careful about wind erosion. As soon as the crops were off in the fall Don planted winter cover crops of rye or ryegrass to keep the soil from blowing, and he had also planted 3,000 feet of windbreaks. He had given up sweet potatoes because they required so much labor.

"Labor is always a headache," Don said. "Some farmers have a house and permanent help, but I prefer to hire temporary help when I need it. I need no labor at all for two months of the year, but harvest time from mid-June to mid-August requires at least two steady men, and sometimes up to half a dozen. Each morning I go down to The Corner [Isabella and Lake streets] in Salisbury to hire workers. I take them back in the evening, and pay them cash each day. The going rate is $1 an hour. We have quite a bit of migrant labor in this area, mostly blacks from Florida or Jamaica. They come to pick strawberries in May, and some stay on until sweet potatoes are finished in mid-November, but most of them go on to New Jersey after the snap-bean harvest in June, then come back for tomatoes in late July and stay through sweet potatoes."

Don sold his truck crops to a local broker or through the farmers' auction market in Salisbury. The auction was important, he said, because it set the price. A lot of the buyers at the auction were local brokers who packed and handled produce for grocery chains, but some were truckers who came directly from Philadelphia, New York, Cleveland, Buffalo, or other cities.

"Mildred and I have lived here since 1946," Don told me. "Her father was a truck farmer in Connecticut, but he moved here when his land was bought for a reservoir. We came here for a visit after I got out of the army and decided to buy this farm. I own 87 acres. Thirty acres was tillable when I bought it, and I have bulldozed and cleaned up ten more acres since then, but there's not much left that's worth clearing. Some of it is sandy knolls that are too dry and the back part is too swampy, but that makes it easy for me to irrigate the tillable land. I use a gasoline motor to pump from a pond that I just scooped out of the swamp with a dragline. I can easily irrigate 20 acres. I also rent 110 acres for $6 to $10 an acre. Ten years ago a man offered me 16 acres of adjoining land for $75 an acre. I sure wish I'd found the money to buy it, because last year it sold for $500 an acre."

Don told me that you can only grow watermelons on the same ground once every eight to ten years, which was one reason for renting land. All truck crops do better on new ground, and he tried to rotate his around, but he also tried to keep them on his own land where he could irrigate them. He used the rented land for 30 acres of corn and 80 acres of soybeans, which were less of a gamble than truck crops. They were grown by machinery and required no outside labor. He got about 40 bushels of soybeans to the acre, 75 bushels of corn. He sold his corn and soybeans to a local feed mill that used them to make broiler feed. He got good prices for them because the Eastern Shore was a major broiler-producing area, and it had a big feed deficit. A lot of corn and soybeans for broiler feed had to be shipped in from the Midwest.

When Don started farming he had 2,000 laying hens for eggs, but he had shifted to broilers and was fattening flocks of 10,500 birds in nine and a half weeks. He was on his own at first, but the price fluctuates so much that it was safer to go with a feed company in a system called vertical integration. Don provided the building, equipment, labor, and electricity; the company furnished the day-old chicks, feed, medication, and a regular service man, and it controlled the marketing. Don was guaranteed $60 per thousand birds plus half the profit, if any, after all expenses had been paid. Some farmers did not like being bossed around by a man from the feed company on their own farms. They grumbled that it was no better than sharecropping, but they liked the steady, year-round employment and income. The manure from the poultry houses was also valuable fertilizer. Don put it on his truck crops and did not have to use any additional nitrogen.

In 1959 Don Wilber was using mainly two-row farm equipment, but he figured that he would have to switch to four-row equipment soon. He would need bigger tractors, but four-row equipment can handle twice as much land in the same amount of time as two-row equipment, with no increase in labor. The usual size of farms in the area was about 60 acres, but Don guessed that they were going to have to get a lot bigger.

How good a prophet was Don? In 1982 when I went back to visit his farm, he laughed about his prediction. "Alan has joined me," he said, "and together we're farming 800 acres, mostly soybeans, barley, wheat, and corn. We have five tractors and six-row equipment. We've gotten even bigger than I thought we would. I never did like the headaches of having to hire workers. I wanted to cut back on labor, so I gradually rented more land, expanded field crop production, and cut back on truck crops. We still have to hire two tractor drivers in the busy season, but that's not quite the same as having to hire field hands. I really do like raising vegetables, but we probably are better off raising soybeans and grain."

Don still had three broiler houses, but he was feeding out 28,000 birds in only seven and a fraction weeks, and he told me that a modern poultry house should hold at least 24,000 to 25,000 birds. A new house would cost around $4 per bird, with three-quarters of a square foot per bird. It gets pretty crowded the last week or so. The price is based on a very complex formula. Don said, "There's not enough real profit in broilers, and most chicken farmers need additional income. I think our own future lies in growing grain to sell to the feed companies because this is still a feed-deficit area, and the feed companies pay good prices."

He turned the farm into a partnership when Alan joined him after he graduated from the University of Delaware in 1980 with a degree in agronomy. They rent small, irregular fields, leftovers from the truck-farming days. They rent farms from five people, plus odd fields from seven others (Figure 11–5). One of the rented farms is seven miles away, and the rest are in between. They pay cash rent of $45 to $50 an acre. They are farming 800 acres, and Alan's goal is a thousand.

They have 350 acres of full-season soybeans, 330 acres of barley or wheat, followed by double-crop soybeans, and 120 acres of corn. They rotate corn, wheat, and soybeans on the heavier ground. Their corn yields average 100 bushels an acre, but even their heaviest land really is too sandy for corn. They irrigate as much corn as they can with a traveling gun irrigation system that was designed for vegetables. It is highly mobile and ideal for vegetables, but in corn they have to check it every hour to be sure that it is working properly.

On sandy ground they double-crop beans and barley. In October they combine full-season soybeans and plant barley in the bean stubble. They harvest the barley early the next June and plant a short-season variety of soybeans in the barley stubble. They harvest those soybeans in late fall, so they are getting two crops of soybeans plus a crop of barley from the field in only two years. They rest it over the winter, plant a full-season variety of soybeans on it the next spring, and start the cycle over again in the second fall. They have been growing mostly barley, but the economics seem to be working toward wheat.

"We sell all our crops to Perdue's feed mill," Alan said. "You probably have seen him on television, Frank Perdue, the one that does commercials for his own chickens. He is one of the biggest broiler producers in the country. He grew up right here in Salisbury, and he still lives here. At his mill they make our corn and beans into feed for livestock, mostly broilers. They ship our wheat to Norfolk to

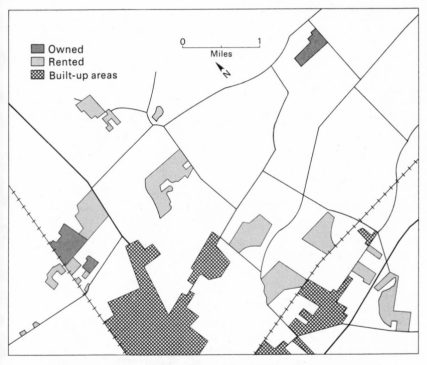

Figure 11–5. The Wilber farm. The Wilber farm consists of many small parcels scattered through the built-up area northeast of Salisbury, Maryland. The Wilbers own two farms and rent five others, plus various odd fields. They must be sensitive to the complaints of their urban neighbors, no matter how unrealistic these complaints may be.

be exported and our barley to Baltimore to be made into beer."

The highly efficient, vertically integrated broiler business on the Eastern Shore has been a model for the expansion of broiler production into many other parts of the South. At the end of World War II northern Georgia and northwestern Arkansas were the only other significant broiler-producing areas in the region, but by 1982 the map of broiler areas in the South looked like it had suffered an attack of measles.

12

They Used to Call
It the Cotton Belt

The South is different. Everyone knows that the South is not like the rest of the United States, but many Americans seem content to criticize and ridicule the differences. They apparently are unable or unwilling to make the effort necessary to understand and appreciate why these differences must be. The South is a tough place to try to make a living by farming because its subtropical climate has produced soils that are seriously deficient in essential plant nutrients. A single crop dominated the region for more than a century and a half, and for many people the South is still synonymous with the Cotton Belt. In fact, the Cotton Belt exists only in history, although it has left clear traces on the landscape and in the minds and attitudes of its people.

In 1950 the U.S. Department of Agriculture published a still widely used map that showed a solid Cotton Belt stretching right across the South from New Mexico to North Carolina. Even then the extent of the Cotton Belt was exaggerated because even at its peak the region had holes in it, and a third of a century later the once-proud Cotton Belt has been reduced to a few tattered remnants.

The traditional Cotton Belt extended from the Balcones Escarpment of Texas to southern North Carolina (Figure 12–1). It was closely associated with the flat to gently rolling topography of the Plainsland South, and it did not extend very far into the uplands of Appalachia and the Ozarks (Figure 12–2). It had a high percentage of black people

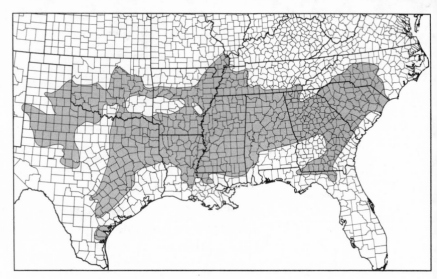

Figure 12–1. The old Cotton Belt. Once upon a time the Cotton Belt extended from New Mexico to North Carolina, but the north-south Balcones Escarpment in east central Texas was generally recognized as the western edge of the South.

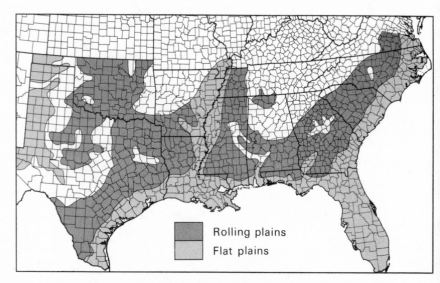

Rolling plains
Flat plains

Figure 12–2. Landforms of the plainsland South. The traditional Cotton Belt was associated with areas of level to rolling topography.

(Figure 12–3). It enjoyed the advantages and drawbacks of a subtropical climate.

Winters in the South are short and mild, with occasional brief spells of severe weather. The January temperature averages between 40°F and 50°F. Even the coldest nights seldom drop below 10°F, though most places have recorded subzero temperatures at one time or another. Frost is rare between the middle of March and the first of November (Figure 12–4).

Spring is long and delightful, but by June the weather becomes oppressive. The heat debilitates and stultifies. It is relentless, merciless, endless. Day after summer day the sun hangs in the sky like a great glob of molten copper, heat shimmers up from the sun-scorched earth, and distant objects dance in the haze. The heat fries your brain and saps your energy. The air is so muggy that even the slightest exertion leaves you dripping with sweat. Nightfall brings no relief because the humid air traps the heat like a heavy, wet, woolen blanket.

The summer sky is flecked with fleecy white clouds, like giant bolls of cotton. They provide welcome shade from the scorching sun, but by late afternoon they can boil up into angry thunderheads that unleash torrential showers. A single summer thunderstorm in the South can smite the earth with twice the rain that many parts of the United States receive in an entire month, and in autumn the rain of tropical hurricanes seems to be coming straight out of a fire hose (Figure 12–5). The South is one of the wetter parts of the world, but its rainfall is both spotty and unpredictable. Farmers expect two or three weeks without a drop of rain even in normal years, and in bad years severe droughts can last all summer long.

Long, hot summers with heavy precipitation provide ideal growing conditions for most plants, but there is a trade-off between climate and soil; the areas that have the

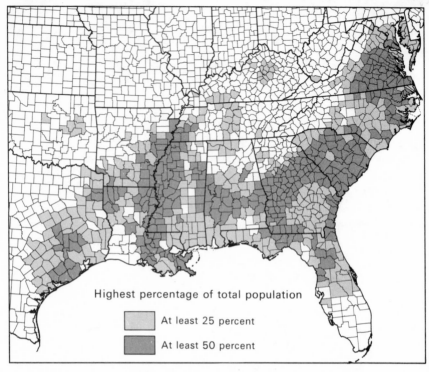

Figure 12-3. Black population of the South. Black people comprised more than a quarter of the total population at some time in virtually all of the counties of the traditional Cotton Belt. In many counties more than half of the population was black.

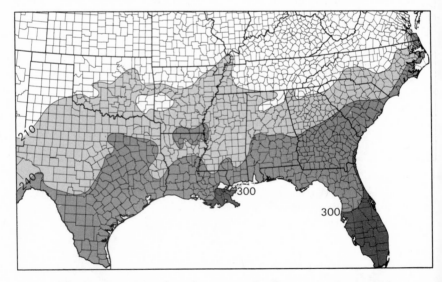

Figure 12-4. Frostfree days. Virtually the entire Cotton Belt had a growing season of seven months or more.

most genial climates often have the most impoverished soils. With a few notable exceptions, the soils of the South were no better than mediocre to begin with, and the cultivation of row crops such as cotton, corn, and tobacco allowed torrential rains to strip away the topsoil and carve gullies deep into the subsoil.

Heavy rainfall leaches plant nutrients from the soil and leaves only the insoluble compounds of iron and aluminum. Percolating ground water concentrates these compounds in the subsoil. They remain soft as long as they are moist, but they cement the soil into an impenetrable hardpan if they dry, especially if they are exposed to the sun. Soil with the rich dark color of humus is rare in the South because hot, moist conditions encourage the bacteria that destroy organic matter. More common is dusty yellow

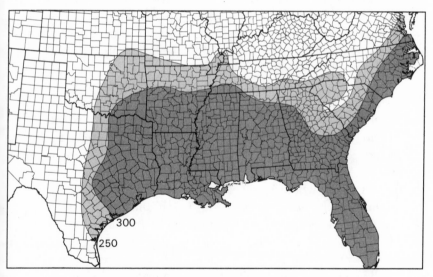

Figure 12–5. Rainfall intensity. The old Cotton Belt is one of the wetter parts of the world. Summer thunderstorms and autumn hurricanes pelt the ground with such intensity that farmers must take special precautions to reduce soil erosion.

sand or bright red clay that sticks tenaciously when it is wet and turns to brick when it dries.

The Englishmen who settled at Jamestown in 1607 discovered that they could not transplant traditional English field agriculture to the semitropical environment, but they needed some commodity they could ship to England to pay for the creature comforts and finery they could not make for themselves. They settled on tobacco, which became the first commercial crop in the United States. The large tobacco plantations of Virginia and Maryland were on the western shores of Chesapeake Bay, and each planter shipped his own crop from his own wharf.

Tobacco production required enormous amounts of labor. At first the planters brought indentured servants from England to work in their fields. When the servant had worked out his period of indenture he was free to produce tobacco in competition with his former master. This competition persuaded planters to shift from indentured servants to slaves for their labor supply, but the damage had already been done.

Former indentured servants moved to the back country and became small yeoman farmers. They raised corn and wheat for themselves and their livestock, and a small patch of tobacco for sale. Patches were appropriate for tobacco because one person could handle only an acre or two. Tobacco exhausted the soil, and the farmer cleared a new patch when the old one was worn out. After two or three crops he abandoned the patch and allowed it to grow up in coarse, beige broomsedge and scruffy, old-field pine.

Farmers were constantly clearing new land while they allowed cultivated land to grow up in brush and trees. At any given time more than half the land was wooded, and much of the rest had only a scraggly growth of weeds, brush, and trees. Visitors from extratropical areas often comment on the amount of "abandoned" land because they do not understand that brush fallowing is an effective technique for maintaining production from semitropical soils. Tobacco has remained an important cash crop for more than 350 years in parts of eastern Virginia and Maryland that visitors thought were "worn-out and abandoned."

The tobacco plantations of Virginia were not the model for cotton plantations. The term "plantation," which has been highly romanticized by some people in the South, originally meant no more than a clearing in a tract of wooded land when it was first used in Virginia. The model for the aristocratic plantation tradition of the cotton South blossomed on the coast of South Carolina, whither it was imported from the sugar islands of the West Indies. The principal early plantation crops of the Carolina coast were rice and indigo. By 1690 planters were growing rice on reclaimed swamps along the major tidal rivers near Charleston. The rice plantations used gangs of workers with primitive hand tools to build dikes and cultivate the crop. The work was nasty because the swamps were hot, sweaty, muddy, and riddled with mosquitos carrying malaria. The Carolina rice planters followed the lead of sugar planters on the Caribbean islands and imported slaves from Africa. The rice planters learned how to manage gangs of slaves on large landholdings, and they created the model that was emulated in other parts of the South with other crops. After 1740 the plantation model was transferred to the cultivation of indigo, which was fermented to extract a deep blue dye. Indigo was an unpleasant and unhealthy crop that was grown by slaves on large upland landholdings inland from the rice plantations. It enjoyed a brief period of prosperity after the Revolutionary War, but overproduction in other areas depressed the price so severely that most indigo farmers had switched to cotton by 1800.

Cotton was little more than a curiosity in 1790 because separating the seeds from the tenacious fiber was so slow and expensive. The development of an efficient cotton gin in 1793 stimulated rapid expansion of cotton production. In 1790 production was negligible. It increased to 50,000 bales in 1800, and it doubled each decade thereafter until the Civil War, when it was more than 4,300,000 bales.

Cotton dominated the South for a century and a half, but cotton cultivation took more than 40 years to spread from South Carolina to Texas. In 1820 most of the crop was grown on the Piedmont of South Carolina and Georgia,

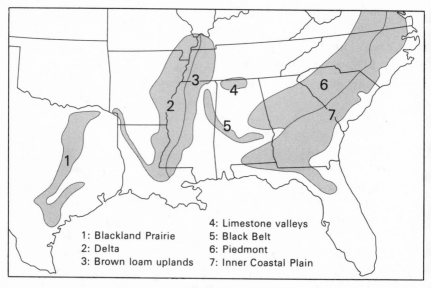

1: Blackland Prairie   4: Limestone valleys
2: Delta   5: Black Belt
3: Brown loam uplands   6: Piedmont
                         7: Inner Coastal Plain

Figure 12–6. Favored areas in the South. Cotton was grown widely throughout the old Cotton Belt, but a considerable proportion of the crop was produced in seven especially favored areas.

and the frontier of white settlement was no farther west than the present site of Atlanta. By 1840 cotton production had moved westward to the limestone valleys of northern Alabama, the Black Belt of Alabama and Mississippi, the Red River Valley of Louisiana, and, most notably, to the brown loam uplands north and south of Natchez and Vicksburg in southwestern Mississippi (Figure 12–6).

By 1860 cotton planters had occupied the more favored areas in the South, except the Blackland Prairie of east central Texas (Figure 12–7). The plantation areas of 1860 still had the largest farms in 1982. Outside the plantation areas farms were smaller and had few or no slaves. Most of these farms depended on a small patch of cotton as their principal source of cash income.

Cotton was easy to grow, but it demanded so much labor that even a large family could handle no more than 10 to 15 acres. The backbreaking tasks of chopping and picking required little skill but tremendous stamina. The semitropical climate that was great for cotton was also great for a whole host of weeds, and all summer long people spent their days in the broiling sun chopping out the weeds with hand hoes. Picking started in August and lasted for a couple of months. Schools closed for a cotton-picking "vacation" so the children could help. The pickers had to bend from the waist to pick the ripened bolls because the cotton plant is a bush that grows only thigh high.

Cotton was grown on farms of all sizes before the Civil War, but the most efficient cotton-producing unit had 900 to 1,000 acres and 60 to 100 slaves. The emancipation of slaves forced planters to reorganize their operations after the Civil War. The planter still owned the land, but he had no one to work it. The freed slaves had no land. This dilemma was resolved by the infamous system known as sharecropping. The planter provided the land, the mules, and the tools. The sharecropper did the work, was charged for half the cost of the seed and fertilizer, and received half the proceeds from the sale of the crop at "settlin' time" in the fall.

The planter divided the large fields of the prewar plantation into sharecropper subunits of 30 to 40 acres. Each subunit had a cabin, a rude unpainted shack cheaply constructed of green lumber with a sheet metal roof. The croppers were more like slaves than tenants because the

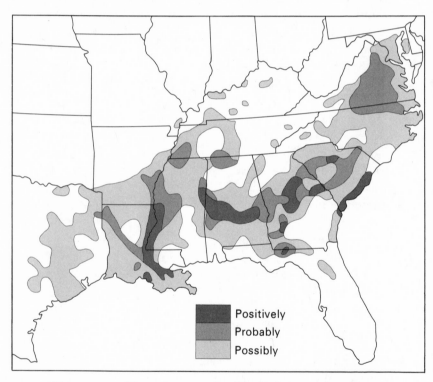

Figure 12–7. Plantation areas in 1860. Defining a plantation as a landholding of more than 500 acres with more than 50 slaves permits identification of the areas where plantations were most important at the outbreak of the Civil War. The plantation areas of 1860 still had the largest farms in 1982.

planter told them when and how to plant, chop, pick, and sell the crop. During the growing season the cropper looked to the planter for "furnish" of food, clothing, and other supplies, which were charged against his share of the crop at settling time. Many planters had their own stores at which they required their croppers to make all their purchases. The store was also the headquarters of the plantation.

Cotton remained king in the South from the Civil War until World War II, but a whole host of problems gradually whittled away its domain. The continuous cultivation of row crops encouraged erosion to claw aching red gullies even on gentle slopes. Many farmers checked gully erosion by constructing low earthen terraces along the contours of their fields to slow down the runoff from heavy rainstorms. Today terraces snake across pastures and through wooded areas as well as across plowed fields. They are a splendid indicator of the amount of land that was cultivated and exhausted in years gone by.

Cotton farmers responded to soil exhaustion by brush fallowing. At any given time more than half of the land in the South was wooded. Most ground probably was used to grow cotton at one time or another, but few counties had as much as a quarter of their area in cotton even at their very peak of production. Cotton farmers also tried to combat soil exhaustion by heavy use of fertilizer. Before World War II farmers in the South used more than half of the nation's commercial fertilizer.

Another problem that afflicted cotton farmers was the boll weevil. Boll weevils sneaked into southern Texas from Mexico in 1894. By 1921 they had invaded the entire South, destroying more than 2 million bales of cotton a year. Farmers eventually learned how to control boll weevils, but much of the country east of the Mississippi River never fully recovered from their depredations.

Cotton farmers were also afflicted by fluctuating prices and overproduction. The only way they knew to maintain their income when the price dropped was to grow more cotton because there were no acceptable alternative crops, but overproduction only depressed the price still more. Furthermore, some farmers kept on growing cotton because it was in their blood. Growing cotton seems to be an incurable disease: once a person has contracted it he can never completely recover. He always has the urge to make just one more crop of cotton.

Before World War I the price of cotton was unsatisfactory, and after the war it got even worse. It tumbled from 35 cents a pound in 1919 to 16 cents in 1920, recovered a bit in the next few years, and then dropped to an all-time low of only 6 cents a pound in 1931. Farmers demanded governmental action. Their demands eventually resulted in the Agricultural Adjustment Act of 1933, which has been the basis for all subsequent farm legislation. This legislation has guaranteed the price of cotton. It has also controlled the acreage that may be planted because farmers would grow entirely too much if the government did not restrict them. Each year the secretary of agriculture determines the total acreage of cotton that may be planted in the United States. This acreage is allotted to the states in proportion to their cotton acreage in a historic base year. The state acreage is allotted to counties and the county acreage to individual farmers on the same basis. The acreage allotment of an individual farmer is the acreage for whose production he is guaranteed a price. He is assessed a severe penalty if he exceeds his allotment.

Government price-support programs helped cotton farmers for a little while, but acreage controls eventually forced many of them out of business. They concentrated their allotted acreage on their best land, cultivated it intensively, and poured on the fertilizer. They doubled their yields per acre between 1930 and 1960, and increased cotton production faster than the secretary of agriculture could reduce the acreage base. Each increase in production necessitated another reduction in acreage, and an ever-larger number of cotton farmers found themselves with impossibly small acreage allotments.

The government price-support program also hurt cotton farmers in the South by encouraging competition from other areas. In 1930 the United States produced more than half of the world's cotton, it dominated the world cotton export market, and the domestic price of cotton pretty well determined the world price. The high support price of cotton became a "price umbrella" that encouraged producers in other countries to increase their own production, and they were able to take over part of the world export market. By 1980 the U.S. share of world cotton production had dropped from more than half to less than a sixth.

Cotton farmers in the South also had to face competition from producers in other parts of the United States, who began to grow cotton when acreage-control programs were temporarily suspended, as they were during World War II. Highly mechanized farms on irrigated land in the West can grow cotton at half the cost in the South, and they can expect two bales to the acre, whereas one is considered good in the South. Cotton has gone West in the United States. In 1910 the three states of Alabama, Georgia, and South Carolina grew 41 percent of the nation's cotton, but by 1982 they had dropped to a mere 5 percent, and California alone produced five times as much cotton as these three states combined.

All cotton farmers have had to face competition from synthetic fibers, such as rayon and polyester. The ratio of cotton to synthetic fibers in textile mills was reversed from 75/25 in 1950 to 25/75 in 1982, and per capita consumption of cotton dropped from 25 pounds in 1966 to only 14 pounds in 1982. Most Americans would be arrested for indecent exposure if they suddenly were stripped of all their clothing that was not made of cotton.

Soil depletion and erosion, boll weevils, discouraging prices, acreage-control programs, competition from other areas and from synthetic fibers, all had buffeted cotton production in the South, but its death knell was not sounded until World War II, when cotton planters lost their labor supply and were forced to mechanize their operations. Large numbers of black people had first started to leave the South during World War I, when northern companies were cut off from their supply of unskilled workers from southern and eastern Europe. During World War II the exodus of black people to northern cities swelled into a flood, and many white workers also left the land for jobs in the new war plants that were being built in towns and cities in the South.

Cotton farmers had been slow to adopt new machinery and labor-saving methods because they had been able to rely on an abundant supply of black and white sharecroppers for the demanding chores of chopping and picking cotton. They were forced to replace hand labor with a new technology when their workers left. They used herbicides instead of hoes to get rid of weeds, and they used mechanical pickers instead of strong arms and aching backs to harvest the crop. They even pressed aircraft into service. In cotton country the summer sky is full of low-flying planes spraying various chemicals on the growing crop. In 1982 the town of Clarksdale, Mississippi, alone boasted 14 crop-dusting services that operated more than 100 planes.

The new technology required large farms, large fields, and large expanses of level land. The eastern South lagged in adopting it because many farms were not large enough to afford it, much of the topography was too choppy, and many fields were too small to use it efficiently. The share of the nation's cotton crop harvested mechanically increased from only 8 percent in 1950 to more than 50 percent in 1960, but even in 1960 less than 10 percent of the cotton crop east of the Mississippi River was picked by machine.

The new technology put small cotton farms out of business. One worker with machinery could handle 100 acres of cotton and produce 100 bales or more, but a family with mules could not make more than about ten bales a year, and they simply could not live on that. In 1949 more than 700,000 farmers in the six states of the old South grew cotton. By 1964 the number of cotton farms had dropped to 180,000, and by 1982 it had fallen to a mere 11,000 (Table 12–1). Cotton had changed from a patch crop to a field crop. In 1949 the average cotton farmer in the old South grew 15 acres of cotton, but by 1982 he was cultivating 230.

Small farmers on marginal land simply stopped trying to grow cotton. In its peak year almost every county from western Texas to southside Virginia had at least 2 percent of its land in cotton, and most had more than 10 percent. By 1982, a mere handful of counties in the South had as much as 10 percent of their land in cotton, and the former

	Number of Farmers			Average Acreage		
	1949	1964	1982	1949	1964	1982
S. Carolina	93,326	25,928	417	12.8	20.32	28.8
Georgia	110,355	22,941	770	14.1	26.9	170.5
Alabama	145,484	42,024	1,458	12.7	19.1	202.2
Mississippi	190,732	50,796	3,710	14.5	28.5	263.7
Arkansas	100,234	21,705	2,019	25.7	56.8	200.8
Louisiana	64,097	16,184	2,371	14.3	31.6	237.3
Total	704,228	179,578	10,745	15.4	28.6	229.7

Table 12–1   Number of Cotton Farmers and Average Cotton Acreage Per Farm, 1949, 1964, and 1982

Cotton Belt had shriveled to a mere skeleton of what it once had been (Figure 12–8). In 1982 the Mississippi alluvial plain and the brown loam uplands just to the east were the last remaining toehold of cotton in the South. Only a few farmers were still trying to grow the crop in other traditional cotton-growing areas (Figure 12–9).

The areas that are still producing cotton are among the most favored environmentally for cotton production, and they have also managed to retain the necessary infrastructure. A farmer cannot grow cotton unless a gin is convenient, but a ginner cannot remain in business unless the local farmers are growing enough cotton to sustain his operation. Many farmers who once grew cotton could not go

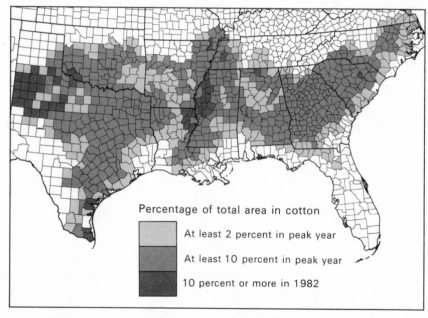

Percentage of total area in cotton

At least 2 percent in peak year

At least 10 percent in peak year

10 percent or more in 1982

Figure 12–8. The shrunken Cotton Belt. Cotton once was king from New Mexico to North Carolina, but by 1982 the old Cotton Belt had shriveled to a pathetic remnant of what it formerly had been.

back now to the crop even if they wanted to because there is no place they could get it ginned. The door is slammed on any prospect for future cotton production in an area when its last gin closes down.

Since World War II the acreage of soybeans in the South has been increasing at the expense of cotton and corn, and soybeans have replaced cotton as the leading crop of the region (Figure 12–10). Soybean production is concentrated in level areas where the land is suitable for large fields and modern machinery because they produce only a modest return per acre, and a farmer must plant a substantial acreage of soybeans in order to generate a reasonable return (Figure 12–11).

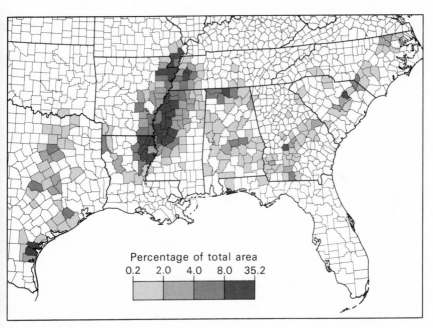

Figure 12–9. Cotton acreage as a percentage of total area in 1982. Cotton remains a major crop in the Delta, but it has virtually disappeared from the rest of the South.

Initially farmers grew soybeans as a fill-in crop when government programs restricted their acreage of other crops, and soybeans still remain a secondary crop in the minds of many farmers. They have made money when other crops have not, but few farmers have developed the affection for soybeans that they have for cotton, and few boast about being soybean farmers. The newest major crop in the South is wheat, which increased tenfold, from 0.5 million acres to 5 million between 1978 and 1982. Farmers started to grow wheat on a large scale when they realized that they could double-crop winter wheat with

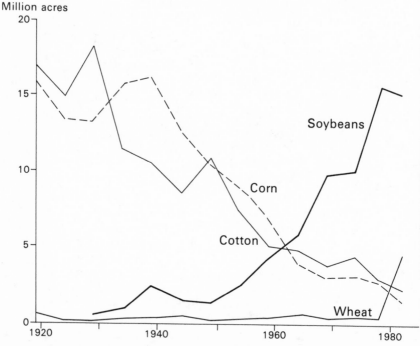

Figure 12–10. Acreage of principal crops in the plainsland South, 1920 to 1982. The acreage of soybeans has been increasing, while the acreage of cotton and corn has been decreasing.

soybeans to produce three crops in two years. Wheat is grown in combination with soybeans throughout the South, except on the Blackland Prairie of Texas at the dry western margin where it is grown alone as a principal crop. The land of cotton has become the land of double-crop soybeans and winter wheat.

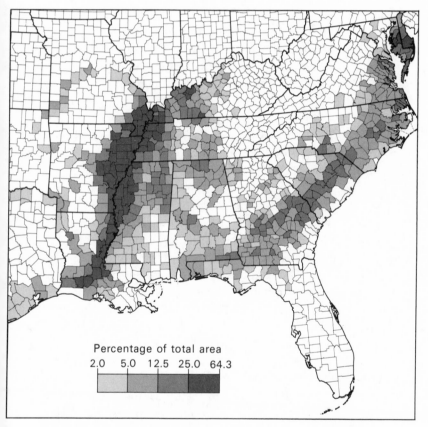

Percentage of total area
2.0   5.0   12.5   25.0   64.3

Figure 12–11.   Soybean acreage as a percentage of total area in 1982. Soybeans are concentrated in level areas that are suitable for large fields and modern machinery.

13

Still Hanging On

Lee County, on the Inner Coastal Plain of South Carolina, is one of the few areas east of Mississippi where you could still find fields of cotton in 1983. The land is flat, with sandy soil the color of bleached khaki. Sheets of standing water glistened in many fields despite the open drainage ditches that slice across the countryside. Every vista is framed by swamp hardwood trees, some clumped in the shallow, oval depressions of all sizes that dot the uplands, some half submerged on the bottomlands along the sluggish streams. Buildings are scattered all over the place: tired and sagging mule barns of weathered, unpainted wood with metal roofs rusting red or black, squat cubes of flue-cured tobacco barns, and houses of all sizes in every conceivable state of repair and occupance.

The D. D. Grant and Son store is a neat, white frame building that dominates Cypress Crossroads. Two modern gas pumps are in front of the store, the porch shades a well-worn wooden bench, and the windows have metal bars to discourage attempts at breaking and entering. A hundred yards down the road is a cotton gin, a great, hulking, nondescript box of rust-blackened corrugated metal, with shed roofs tilted out over the unloading areas. Behind the store are various white, wooden buildings for storage, three cylindrical metal grain bins, and open sheds with metal roofs that shelter cotton pickers, sprayers, cultivating equipment of all kinds, and a veritable army of tractors.

Don Grant, 56, is one of the few remaining cotton farmers in South Carolina. In 1949 nearly 100,000 farmers in the state grew cotton, but in 1982 they numbered fewer than 500, and Don alone grew more than 1 percent of the total acreage of cotton grown in the entire state. "I'm farming 2,000 acres," he said, "half in cotton and half in soybeans. We own about 60 percent and lease the rest. My mother's father farmed here. He built this store to serve the people who worked for him. The farm office is in the back of the store. My father came here as a schoolteacher and married my mother. He added to his farm, and so have I.

"In 1943 this farm had one tractor with a disc harrow, and everything else was mules. I couldn't even begin to tell you how many tenant houses there were. Seems like every time you turned a corner there'd be a house with a family in it. Back in the woods every opening is an old house spot. Some were sharecroppers and some were day laborers. It depended on the ages and number of children, but you could figure that one family could handle about 15 acres of cotton. Then each mule needed two acres of corn and two acres of oats, and you needed some corn for the hogs and the family. I'd like to tear down some of these old houses, but I can't do it while the old people are still living in them. They cost me money for taxes and upkeep, and I can't afford to fix them up as nice as I would like them to be."

In 1947 he got a second tractor and a turning plow, and he has just grown with machinery ever since. He has three two-row cotton pickers, two ground sprayers, a six-row cotton planter, an eight-row bean planter, and about ten tractors. He said that's more than he needs, but they are used for all sorts of odd jobs. A farmer needs 150 acres of cotton to support a two-row picker, and it costs about $80,000.

Seven men work on the farm, and two help around the

store and the gin. Three are white and six are black. "They are all my friends," said Don, "but we don't socialize." All but two have free housing. Every one of them has worked for Don for at least 15 years. They were a proud and loyal lot. Each one wore clean and neatly pressed coveralls or a light blue shirt with dark blue trousers. His first name was on a patch sewed on his right chest. Any filling station in the country would be happy to have such a trim and tidy crew.

"Cotton is our bread and butter," Don said. "We have 1,000 acres, and we get 800 to 850 pounds per acre. You need at least 600 pounds to make any money. I wouldn't want to do without soybeans, but I just love cotton. It's been good to us. The challenge is growing the crop, and it's just not as much fun once it's been picked. My grandfather and father bought an existing gin in the 1930s, and we have improved and modernized it through the years. I could do without it because there's another one half a mile down the road. We sell our cotton through cotton merchants. We need to pay more attention to marketing, as much as we pay to production, because a cent a pound can make a lot of difference when you sell 1,500 500-pound bales.

"We have 1,000 acres of soybeans and get 25 bushels to the acre. They came in during the 1950s. They were a good money crop, simple and not too expensive to grow, low labor cost, and readily marketable. They replaced the corn and oats we had grown to feed the mules. Some farmers with small acreages of cotton and tobacco have switched completely to soybeans. They are a good crop for a factory worker who is farming part-time on the side. Livestock? I don't want anything on the place that I have to feed. I'm a cotton farmer, and I don't have any need for livestock."

Lee County, South Carolina, is one of the few remaining "islands" of cotton production east of the great alluvial

plain of the Mississippi River, which is more commonly
known as the Delta. The Delta is monotonously flat and
poorly drained. The dark soils are rich in organic matter,
and they are extremely fertile when they are drained and
protected against the annual spring rampages of the river.
The Delta is one of the few areas in the United States
where land has been cleared and placed in cultivation
since World War II. More than 3.5 million acres of fine
bottomland hardwood forest in Arkansas, Louisiana, and
Mississippi were converted to farmland between 1945 and
1965.

George Baird's farm is in the very heart of the Delta in
Sunflower County, Mississippi. The entire county has a
slope of less than a foot to the mile. His farmstead is on the
south side of Mound Bayou three miles north of the village
of Inverness. The modern farmhouse is a handsome, red
brick rambler that is shaded by grand old trees. East of the
house are an old pecan grove, a new fruit orchard, and a
grape arbor. To the west are two large, open sheds for
machinery and various smaller buildings. Tractors, plows,
cotton wagons, and other equipment are parked all over
the place. Down the gravel road is a row of four attractive,
one-story, red brick houses with white-painted, decorative
iron grillwork. They would be at home in any middle-in-
come suburb. Several late-model cars are parked in front
of each house.

George Baird, 68, grew 950 acres of cotton in 1982. His
grandfather had bought and cleared the land in the 1880s.
His father had farmed with mules. They once had 27
sharecropper families, but the last one left when the
Bairds converted to machines in 1951. George has four
tractor drivers, three full-time and one old man on Social
Security who works part-time. George pays them by the
hour and gives them free housing down the gravel road
past his farm buildings. His youngest son manages the

farm under his supervision. He had expected one of his two older sons to join him on the farm, but they decided to go off on their own. George cosigned their notes to help them get started.

"First of April we subsoil, plow, and fertilize," George said. "We plant at the end of April or first of May. After the plants emerge we cultivate three to five times for grass control. The poisoning season starts the first of July, and you spray for whatever insects are out there, bugs, lice, thrips, cutworms. We apply most chemicals by airplane because it's too wet for ground machines. My chemical bill runs better than $50 an acre.

"We defoliate between the tenth of September and the first of October when the top bolls you plan to harvest are too hard to cut with a sharp knife. We 'foliate by plane. It takes three to five days for the leaves to begin to drop, and we start picking ten days after we 'foliate. Last year was a good year, and I got 800 pounds of lint cotton to the acre. I shoot for 750. After picking we cut stalks, subsoil, plant a cover crop of vetch, and plow it under first of April next spring.

"Last year I had 120 acres of soybeans on land I didn't want to put in cotton. Most farmers run around 65 percent cotton and 35 percent soybeans, but I can't seem to make any money on beans. I get 25 to 30 bushels to the acre. A lot of land around here is going into rice. I haven't tried it because I'm making money on cotton. I've thought about diversification, but then I figured I could do better if I settled on one thing. It would cost me at least $250,000 dollars for equipment if I went into rice, and I already have around $350,000 worth of cotton equipment."

Cotton is still hanging on, but just barely, in the subhumid Blackland Prairie of east-central Texas. The Blackland Prairie is at the foot of the Balcones Escarpment, where the South ends and the West begins. Its name comes from the rich, milk-chocolate color of its deep fertile soils. The level topography is well suited to large machines. Many fields are unfenced, and the rows end abruptly at the highway right-of-way. Wooded areas are mainly near streams. The scarcity of trees on the uplands hints that the climate is marginal for agriculture. The average annual rainfall of 30 to 35 inches is erratic and unpredictable. Summers are so hot that water loss by evapotranspiration runs neck and neck with precipitation. The July temperature averages 85° F, ten degrees hotter than in Iowa.

The Blackland Prairie was settled after the Civil War by people whose small farms worked with family labor turned the Blackland into a major cotton-producing district. The contemporary landscape hints that the agricultural economy has changed. Numerous abandoned farmhouses say that farms have been enlarged. Working farmsteads and implement dealers still find space for cotton-picking machines and wagons, but combines, grain drills, and metal grain bins are elbowing them aside. The cotton gins in the small towns look old and tired and rusty, but the grain elevators are spanking new.

The Blackland Prairie is at the margin of the area where cotton can be grown without irrigation, and cotton has given way to less thirsty crops, such as grain sorghum and winter wheat. Sorghum replaced corn in the early 1950s, when cotton farmers switched to machines and no longer needed corn for their mules. Sorghum is a better cash crop than corn in an area with only limited rainfall. In the late 1970s dry years and changing relative prices gave wheat an advantage.

Wilbert Vorwerk's farm is 25 miles northeast of Austin in eastern Williamson County, the leading cotton county of the Blackland Prairie, which had 47,358 acres of cotton in 1982. In 1949 Williamson County alone grew more cotton than the entire Blackland Prairie was growing in 1982, and in 1924 the county had an astonishing 318,000 acres—44 percent of its entire area—in cotton. Much of the crop was grown in the hilly western part of the county on steep slopes with thin soils that probably should never have been cultivated.

Wilbert's farm is immaculate. The modern, one-story, tan brick farmhouse is shaded by pecan trees. It sits on a spacious lawn behind colorful flower beds and carefully trimmed shrubbery. Back of the house is a cavernous metal shed for machinery. Twelve cotton wagons with wire mesh sides were lined up in two neat ranks between the shed and the highway. Long straight rows of cotton and sorghum, completely free of weeds, marched across the black soil of the flat plains toward the horizon, with no sign of fence or tree. The whole operation reminded me of a cash-grain farm on the prairies of Iowa or Illinois, except that the crops were cotton and sorghum rather than corn and soybeans.

Wilbert, 43, owns 277 acres and rents 510 more in four farms that once supported four families. He has one hired hand and hires a local Spanish family to chop weeds if herbicides fail to get rid of them. He said that a family could only handle 40 acres of cotton back in the mule days, but that a family would need at least 400 acres to make a living farming in the area today, and he realizes that he is going to be forced to get bigger.

All of his land is in crops. He usually grows half grain sorghum and half cotton, but he has gone as high as two-thirds cotton when the price was right. The usual rotation in the area is two years of grain and one of cotton. He said that cotton is more work than sorghum, it takes longer, and it is more of a gamble, but in a good year it makes a lot more money. He hopes for a bale to the acre but averages around 400 pounds, which is pretty good for a subhumid area. His sorghum yields 4,000 to 4,500 pounds per acre, well above the county average of 3,000. He maintains crop trial plots for the county extension service and various seed companies. The seed is free, and he has a chance to see how the different varieties will do on his own land.

Wilbert may have to spray six to eight times if insects are bad, but he is trying to encourage natural predators to avoid using chemicals. He sets pheromone traps to catch and count insects so he knows when to spray. "Pheromones? They are chemically synthesized female sex scents that attract males." He is constantly experimenting with new ideas. He has tried corn, but the area does not get enough rain to grow good crops. It is too hot and too dry for soybeans. Some local farmers started to grow wheat when the prices were good, but both yields and prices have dropped, so he is glad he didn't try.

Maurice Clack, on the other hand, has switched over completely from cotton to wheat. His farm is on the Blackland Prairie 40 miles northeast of Dallas. The northern Blackland is more dissected than the central and southern parts. Broad, shallow, wooded valleys separate gently rolling uplands that are treeless and windswept. Some of the steeper slopes still have low earthen terraces that were built for erosion control in the cotton days. Cattle are everywhere because every farmer seems to have a small herd of 10 to 40 cows, "just to clean up the grass along the creek," but few farmers depend on cattle for their livelihood.

Maurice, 56, owns 235 acres and has pulled together an operation that is large enough by renting 1,500 acres from 12 to 13 separate landlords. Each parcel was a separate cotton farm back in the mule days. He has so many, he said, that every now and then he will be headed home and suddenly realize that he forgot to plow one. Cotton was his main crop before 1980, and he grew 600 to 700 acres, but insects, weather, and prices all were so terrible that he switched to wheat. He still has all of his cotton equipment, and he might go back if the price got high enough, but he is not sure where he could get it ginned because the number of cotton gins in the county has dropped from twelve to two in ten years.

"Wheat really took off in this area in 1976," Maurice said, "and it has been a good money crop. Our yields have nearly doubled. In 1976 we were getting 17 to 18 bushels an acre, but now we average 35. We have better varieties and better knowledge of the fertilizers needed. We wouldn't think of not using anhydrous ammonia any more. Wheat is cheaper and a lot less risky to grow than cotton. A fellow could probably sit in his living room and make money growing wheat, just by hiring custom operations by telephone. We can easily handle five, six, seven thousand acres of wheat if it's suitably located, but cotton and row crops require more labor and more capital. The shortage of labor may force us to stick with wheat."

In 1949 the Blackland had more than 2,640,000 acres of cotton, but by 1982 it had dropped to fewer than 200,000 acres. The Blackland Prairie, it seems, has been added to the long list of former cotton-producing districts.

14

Pine Trees,
Pastures, and
Poultry

Farmers in the six states of the old South grew
10.8 million acres of cotton in 1949, but in 1982 they grew
only 2.3 million acres. What have they done with the 8
million acres of land they no longer use for cotton? They
use some of it for other crops, such as soybeans and wheat.
They use some of it for pastures of highly variable quality
that are grazed by cattle of equally variable quality. They
have sold or leased some of it to pulp and paper companies
that produce forest products. But most of it, sad to say,
they have merely left to grow up in broomsedge, weeds,
brush, and scruffy, second-growth woodland. It seems to
serve little function other than to hold the rest together.
Vast areas in the South have become dark, monotonous
seas of pine forest (Figure 14–1). Many highways run for
miles through long stretches of "pine tunnel," broken only
by an occasional cleared patch.

Trees, as far as the cotton farmer was concerned, were
merely weeds that took over fallow fields and had to be
cleared before he could use the land once again to grow
crops, but the wood-using industries have capitalized on
the "bonanza forest," forest that grew naturally on the
abandoned cotton fields of the South. Wood processing has
been a major growth industry in the sometime Cotton
Belt. The number of pulp and paper mills in the region
increased from 66 in 1952 to 121 in 1969, and their average

daily capacity grew from 414 tons in 1952 to 704 tons in 1969 (Figure 14–2).

Some landowners in the South have become keenly interested in forest management, but many were unable or unwilling to pay for good forest management practices after the bonanza forest had been harvested, so the pulp and paper companies have had to buy large acreages of

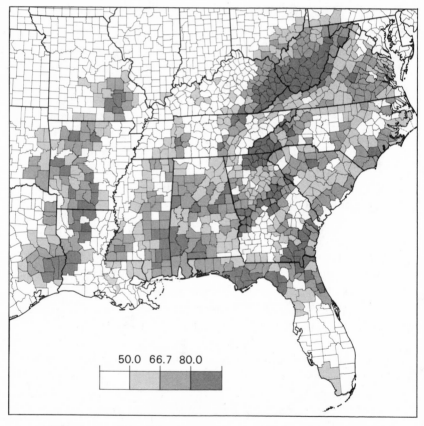

50.0 66.7 80.0

Figure 14–1. Forest land as a percentage of total area around 1982. The South is the most heavily wooded region in the United States.

land to ensure a reliable supply of raw materials. They began to acquire land after World War II, and by 1960 they owned more than 24 million acres, nearly a quarter of all the woodland in the six states of the old cotton South (Figure 14–3). They owned more than half of the land in some counties; the extreme case was Dixie County, in the armpit of Florida, where they owned 82 percent of the entire county.

Most of the former cotton land in the South that is not wooded has been turned into poor pasture for small, uneconomic herds of beef cattle. Beef cattle have been one of the bitter disappointments of the region since World War II. Enthusiasts saw old cotton fields being converted to pastures, and they confidently predicted that the South was destined to become one of the leading beef-producing areas in the nation. The glowing optimism of the early 1950s has given way to a more sober reality. The number of cattle in the six states of the old cotton South jumped from about 5 million before 1940 to about 10 million in 1954, but since then it has wavered between 8 and 10 million.

The greatest concentrations of cattle in 1982 were around the fringes of the old cotton South in Appalachia, the Ozarks, eastern Texas, and Florida (Figure 14–4). There were lesser concentrations within easy driving distance of some metropolitan centers, The map shows few cattle in areas that seem, at least from the highway, to be well stocked. There are three reasons for this apparent discrepancy. First, the principal highways follow the easiest terrain, which is also the area most likely to be cleared and used for pasture, whereas areas back from the highway are more likely to be rougher and more heavily wooded. Second, one's eyes may exaggerate the number of cattle because there is little else to see in the islands of cleared land that occasionally break the long tunnels of

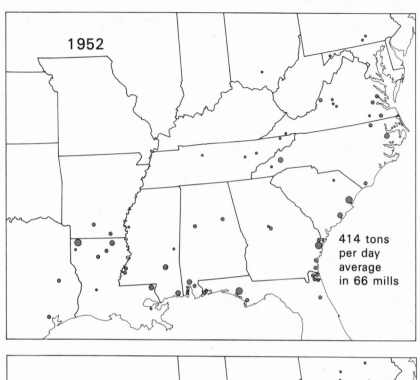

1952

414 tons
per day
average
in 66 mills

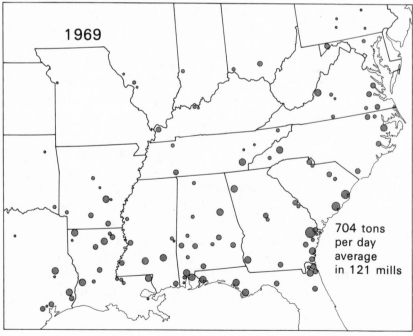

1969

704 tons
per day
average
in 121 mills

Figure 14–2 (*facing page*).   Pulpwood mills in the South in 1952 and in 1969. The number of mills nearly doubled, and the average mill capacity also nearly doubled.

pine trees. Third, cattle herds really are concentrated near the major highways, both for ease of access from the

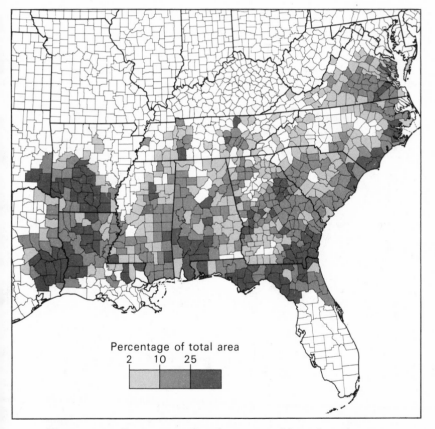

Percentage of total area
2   10   25

Figure 14–3.   Percentage of total area owned by pulp and paper companies and other forest industry companies around 1982. These companies owned nearly a quarter of the woodland in the six states of the old cotton South.

city and for showcasing the animals. Many herds of cattle in the South are status symbols for city people who enjoy playing cowboy on weekends.

Native grasses had colonized abandoned cotton fields, but they were grazed in desultory fashion. Few farmers had much interest in good livestock and pasture management before World War II. The combination of high beef

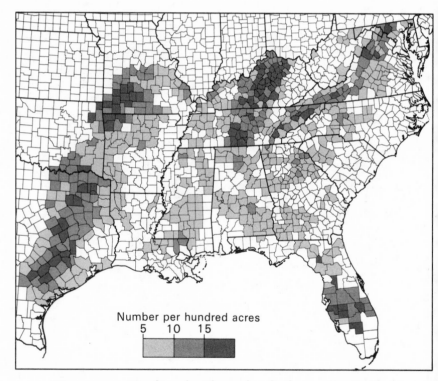

Number per hundred acres
5   10   15

Figure 14-4. Number of cattle per hundred acres in 1982. Despite the exuberance of enthusiasts, few areas in the plainsland South outside the Blackland Prairie of Texas and central Florida had significant numbers of cattle. The principal concentrations were in the western Ozarks, the limestone lowlands of Kentucky and Tennessee, and the Great Valley.

prices and labor shortages during the war stimulated widespread interest in cattle. Enthusiasts noted that some of the plants which had been used to control erosion in the old cotton fields could provide forage, and they made much of the fact that winters were so mild that cattle could graze out of doors throughout the year, thus saving the expense of winter feeding and shelter.

Cattle farms provided both tax relief and visible status symbols for successful business and professional men. They could invest their money profitably, yet still claim heavy tax losses. The cattle farm was a pleasant place to invite friends for the weekend, and it conferred on its owner the privilege of wearing a ten-gallon hat, high-heeled boots with pointed toes, a western shirt, and a bolo tie, which made him the envy of some of his colleagues and all small boys. Unlike commercial farmers, who had to make a living by farming, many city farmers went in for purebred registered livestock.

Many problems have kept cattle from fulfilling the dreams of visionaries. A successful beef operation in the South requires at least 100 cows and 300 acres of good grazing land, but most cotton farms were too small, and most farmers lacked the capital to enlarge and develop them adequately. The agricultural credit system was based on six-month loans for cotton, and bankers were reluctant to make the longer-term loans that were necessary for cattle.

Pasture and grazing land in the South has a remarkable range of quality, but most of it is near the bottom end of the scale. The region has no native forage plants that provide good, year-round grazing, and pasture management required a mix of plants that could survive the heat of summer and the chill of winter. The new pastures on the old cotton fields had to be fenced, and good fences are expensive. The pastures had to be stocked, and some neophytes learned that they were not the world's best judges of cattle. The new pastures also needed a good supply of water because mature beef animals drink ten gallons a day; landowners dammed small streams and speckled the countryside with new stock-watering ponds. Marketing facilities had to be developed; some of the early beef farmers had to truck their cattle out of the region in order to obtain a reasonable price for them.

One of the first livestock areas in the South was the Black Belt of Alabama and Mississippi. Lawrence C. Alsobrook, county agent in Dallas County, Alabama, told me, "The boll weevil hit us here in the Black Belt in 1914. We could no longer grow cotton, many of the black people left, and much of the land lay idle. The native grasses moved in and furnished sod. We turned cattle onto it, but we did not really know much about forage crops and livestock until 1930, when the state established a research station in the western part of the county. Farmers around here thought the people at the research station had gone crazy when they started putting fertilizer on grass.

"A man must have at least 100 beef cows if that's his total livelihood, and a lot of our operators run 500 or more. For each animal unit we like to have an acre of grain, either oats or wheat, that we graze in winter; an acre of permanent pasture; and an acre of Johnson grass for spring grazing and hay. We like to went broke fighting Johnson grass in the cotton fields, but then we learned how to cultivate it, and now it's making money for us. This is a grain-deficit area, and we sell calves grass-fat at around 600 pounds. We prefer a fall-dropped calf. They're large enough to consume a lot of forage during our peak period, from March to May, and we sell them just before our low forage time."

Jimmy Minter, Jr. and his son, Jimmy, III, have a large beef and cotton operation ten miles southeast of Selma at Tyler. "You can't miss it," they told me when I called them in the spring of 1983. "The pavement ends at the store" that serves as headquarters for J. A. Minter & Son. A red dirt road continues eastward, and another winds up the slope to the north. Sturdy, barbed-wire fences line the roads, but nothing separates the pastures from the wooded areas behind them and cattle of many colors wander back and forth.

The store is a hulking white building with a false front. The U.S. Post Office is tucked into a separate room in the corner of the building, and an American flag droops limply from a short pole in front. A corrugated metal awning shades a small paved area at the entrance. The store has no gas pumps, and its shelves are empty, but a comfortable room at the back is the farm office. Behind the store are a wooden shed large enough to hold three cotton pickers, a cluster of metal grain bins, and a dozen tractors. A cotton gin is on the corner diagonal to the store.

Jimmy Minter, Jr. is 78 years old. His ancestors came here about statehood time, in 1819, from North Carolina. They acquired 11,000 acres, and his family has been here ever since. He is the fifth generation. His father started here in 1890, and in 1901 he built the gin. They had 165 to 200 tenants at their peak (Figure 14–5). "I could take you to a hundred places where there used to be tenant houses," he said. "I used to just live on a horse supervising cotton. I still lease a bit of land to some old men who can rent land from me as long as they live. They pay me less than I have to pay when I rent land, and they have to hire most of the work done for them, but they just like to have some cotton."

Jimmy Minter, III, is 42, and Jimmy, IV, is 10. "We stopped leasing and started farming the land ourselves with machinery when I got home from college," he told me. "We had to do it ourselves because we couldn't get tenants. We have nine full-time workers and some part-time labor. All of the workers have free houses on the place. We kind of grew into it, but we're going to get out because it costs us too much. I don't like all of the worries with labor. Every time Willie has a flat tire you've got to stop what you're doing and get it fixed."

The Minters grow 1,700 acres of cotton, average 750 pounds to the acre, and run their own gin. They grow 800 acres of soybeans and get 25 bushels to the acre. "We grow them when there are acreage restrictions on cotton," Jimmy said, "but I don't think much of 'em. They take more manpower and equipment, weed control is harder, the price is bad, and disease is starting to be a problem. They're just not a good crop for this area. I grow wheat if the price looks good enough, but this land is not good for wheat. Corn is also a poor crop here. Cotton really is about the only good crop in this area."

Several thousand acres of their land are wooded. They started a timber program in 1938 when they brought in a federal forester to give them advice. Then he set himself up as a private consultant, and they have been using him ever since. They are culling the hardwoods and trying to work the timberland into pine. They thin for pulpwood at 15 years, sell pulpwood at 25, clear cut for saw logs at 35, and replant in pine.

"I really like the cattle operation," Jimmy said. "In 1928 Dad started a herd of 400 brood cows. We had 1,500 acres of pasture on land that was too wet to cultivate, but disease ran us out of the cow business. We sold the brood cow herd five years ago and went to winter grazing. We buy 400 to 600 calves at 450–500 pounds in the fall and hope to put 300 pounds on them in 150 days.

Figure 14–5. The Minter plantation. The Minter plantation is inside the bend of the Alabama River east of Selma, Alabama. Nearly half of the tenant houses that had stood on the plantation in 1932 were not there fifty years later.

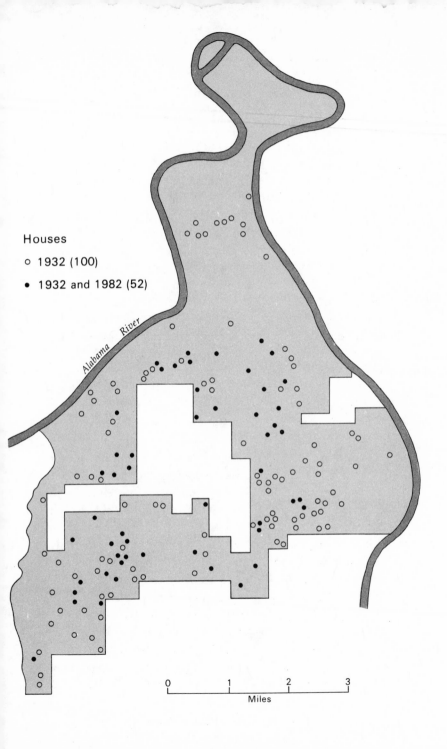

Houses

○ 1932 (100)

● 1932 and 1982 (52)

Alabama River

0   1   2   3

Miles

"The cattle graze from mid-November until May on prepared winter pasture of clover, rye, and ryegrass. We supplement the pasture with some hay, but buy no feed. We sell cattle at an auction at the Holiday Inn in Selma. Yessir, that's right, at the Holiday Inn. The cattlemen in this area publish a map of their farms, with a list of the cattle for sale at each one. The buyers come out to the farms to take a look at the cattle before the sale, and then they bid on 'em by lot number. This is our fifth year. We've already had some good years and some bad ones, but I want to expand the cattle operation. Hogs? People around here won't fool with 'em. 'Are you a gentleman,' they ask, 'or do you raise hogs'?"

Poultry have received far less publicity than cattle, but broilers have been the true livestock success story of the South since World War II. In 1949 farmers in the six states of the old cotton South sold only 95 million broilers, but in 1982 they sold 1,898 million, or more than half of the nation's meat-type chickens. They have changed chicken from a luxury to a staple in the American diet. Once nearly every farm had a small barnyard flock that survived by scavenging, and chicken was reserved for Sunday dinner, but the broiler business has become so highly specialized and efficient that chicken is our cheapest meat.

Specialized poultry production began in the 1930s in "problem farm areas," areas that had large numbers of small farms with a low income per farm and per family. These areas needed a new source of farm income because previous sources—fishing, eggs, and vegetables on the Eastern Shore; apples in northwestern Arkansas; and cotton on the Piedmont of northeastern Georgia—had failed or were no longer adequate. Feed dealers in these areas saw a chance to expand their business by giving farmers chicks and feed on credit, with the loans to be repaid when the birds were sold.

Many small farmers were whipsawed by the volatile market for broilers, however, and after World War II feed companies developed vertical integration. Companies have their own hatcheries, feed mills, and processing plants. They deliver day-old chicks to the farmer, fill his feed bins once a week, guarantee him a price, and collect the fat birds for processing and marketing. The farmer provides only the broiler house and his labor. Companies exercise such close supervision that some farmers find it onerous, but they do not object to the check the company sends them.

The feed companies have encouraged breeders to develop plumper birds that can convert feed faster and more efficiently. Producing a 4 pound broiler took 17 pounds of feed and 15 weeks in 1940, but in 1980 it took only 8 pounds of feed and 7 to 8 weeks. Poultry houses have been enlarged and made more efficient. In 1940 houses that held 1,500 birds were considered big, but modern broiler houses hold 20,000 birds or more. The mechanization of feeding and watering has reduced labor requirements from 250 hours per thousand birds in 1940 to fewer than 25 hours in 1980.

Chickens convert feed more efficiently than cattle or hogs. In 1980 it took eight pounds of corn to produce a pound of beef, four to produce a pound of pork, and only two to produce a pound of chicken. The price of chicken has increased only slightly since World War II, while the price of beef and pork has more than tripled. Per capita consumption of chicken has already passed pork, and soon it will be greater than per capita consumption of beef.

Broilers have been an attractive option for many small farmers, especially those who can find off-farm jobs, because they produce regular income and require only a few hours work each day. Much of the work is light enough to be handled by women and children. The broiler business has been the salvation of many rural areas in the South that otherwise might have had to abandon farming completely. Broiler production is highly concentrated geographically in intensely specialized areas that are widely scattered (Figure 14–6). The modern industry originated on the Eastern Shore before World War II. It grew rapidly in northeastern Georgia and northwestern Arkansas immediately after the war, and subsequently developed elsewhere.

One might argue that broiler production really is manufacturing rather than farming because it has so little tie to the land. The broiler districts are feed-deficit areas, and much of the feed has to be shipped in, mainly from the Midwest. The long, low, one-story broiler houses are essentially factories that use the birds as machines to convert raw materials—corn and soybeans—into a finished product—meat. There are few other industrial processes, however, in which the machines are eaten after they have done their job.

One of the by-products of the growth of the broiler business has been a major waste-disposal problem. Chickens are efficient manure producers; a four-pound bird drops a quarter of a pound a day, and a 20,000-bird house can generate two and half tons daily at the end of the feeding period. This stuff is excellent fertilizer, and ideally it should be returned to the land. Some broiler farms have developed subsidiary beef operations based on pastures enriched by poultry manure, but they need a lot of land to handle the material from 100,000 busy chickens, and they have to pay somebody to haul and spread it, in addition to the expense of machinery, gas, and oil.

In recent years poultry farmers have made a major breakthrough in managing manure: they recycle it by feeding it back to the animals. It is rich in protein, and after it has been dried out and flavored with molasses it looks like soybean meal. It can be fed to chickens or cattle, and the animals do not seem to notice the difference, at least not after they have gotten used to it.

Fillmore Peek, 63, owns a 170-acre broiler and cattle farm on a hilltop eight miles north of Carrollton, Georgia. About 100 acres is open grazing land that shades into scrub hardwood forest on the steeper lower slopes. "My daddy had four sharecroppers on this farm," he told me. "I

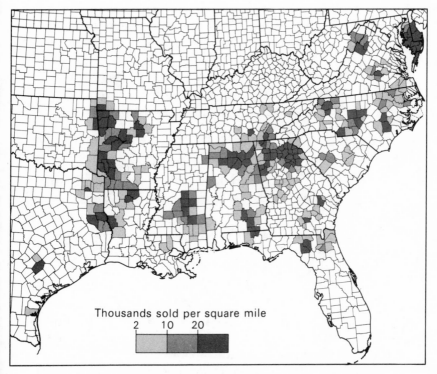

Figure 14–6. Thousands of broilers sold per square mile in 1982. The explosive growth of the broiler business has been the true livestock success story of the South since World War II. It has changed chicken from a luxury reserved for Sunday dinner into our cheapest meat.

started off growing cotton, but I had to stop when the workers discovered that they could get more from the county welfare office than I could afford to pay them to pick it by hand. This land is too steep for machinery. I put it in pasture, but I can't even get help to bale hay.

"I have three poultry houses that cost me $30,000 apiece. I put three batches of 12,000 birds a year through each one. It's a full-time job. I make about $2,000 a batch. I have a breeding herd of 40 cows on the pasture, and I expect each cow to drop a calf each year. I sell the calves at around 400 pounds at the auction in Carrollton. I used to spread the litter from the poultry houses on my pastures. It's real good fertilizer. A few years ago my manure spreader jammed up on me, and it began throwing out the stuff in big chunks. I noticed that the cows were really gobbling it up, so since then I have been mixing in 3 percent corn to make winter feed for the cattle. I have to feed them from December through March."

The broiler-producing district of northwestern Arkansas is in the Springfield Plateau section of the Ozark Uplands. Headward erosion by streams has fretted deep valleys in the edges, but the central strip is a wide, gently rolling upland. The soil is too thin for successful crop cultivation, and most of the land carries verdant pastures that are grazed by beef and dairy cattle.

In 1910 northwestern Arkansas was apple country, but a combination of poor prices and problems with insects and diseases decimated the apple business. Farmers began scratching around for a new source of income. Their farms were too small for beef-cattle operations, and the soil was too thin for crops, but broilers worked well because they required little land or capital. The large, rural labor force in this area has attracted light industries, and today many "farmers" tend broiler houses before and after they commute to factory jobs.

The Dixieland Road north from Rogers, Arkansas, is a four-lane highway that turns into a narrow red dirt road when it crosses Highway 102. A sign beside the road heralds the city limit of Little Flock, population 663, which was incorporated to keep the town of Rogers from putting its dump in the area. Half a mile beyond on the left are five 32-by-400-foot broiler houses set at an angle to the road. The roofs and lower walls are corrugated metal. The upper half of each side wall is a mesh screen with adjustable canvas curtains. Unpaved dirt lanes give access to cylindrical, metal feed bins beside each house.

Forrest Bland and his wife, Ruth, obviously like and respect each other and enjoy doing things together. Forrest, 58, is leathery and lean as a whip. He was slightly suspicious of what I was up to, but he guessed it would be all right if I talked to them while they repaired the feed mechanism in one of their poultry houses. When we entered I gasped and choked at the acrid smell of ammonia—"That's the smell of money," Forrest told me—but after a few minutes I hardly noticed it.

The floor of the broiler house was alive with clusters of baby chicks that had just been delivered, jostling, pushing, shoving, running, falling, cheeping balls of curious yellow fluff. A metal feeder trough ran the length of the floor. An endless chain with flat links rattled along it at a slow set speed. The chain was loaded when it passed beneath a feed bin that was automatically filled from the large bin outside. Down the sides and across the center were 64 eight-foot, metal watering troughs that were filled automatically. Down the center of the house dangled a line of 22 circular, metal "brooder stoves" that burn natural gas for heat in winter. Along either side of the ceiling were rows of 20 bright white lights that were kept on all night to encourage the chickens to eat as much as possible. The interior was a piano-wire jungle. "Everything is on wires,"

Forrest said, "because we have to raise it off the floor with a hand winch before the catchers come to get the chickens. He told me that the current cost of a 16,000-bird house was $48,000. "Ruth and I started off with nothing," he said, "and we have built it all up. I was a tail gunner on a Marine SBD Dauntless dive bomber in the Pacific during World War II. When that old airplane went into its bombing dive I sure wished I was back home in Arkansas! I saved all of my money, and Ruth saved all of hers. She was working as a bookkeeper for an automobile company. After the war was over we bought 40 acres and a herd of milk cows. In 1965 or 1966 I just wasn't getting along as good as I wanted in the dairy business, and I started talking to the integrators. Tyson seemed to offer the best deal, and we've stayed with them ever since.

"We started off with one old broiler house. I milked the cows and Ruth tended the chickens. We've added more houses when we could afford it. We run four to six batches a year in each house. We feed them to 4.5 to 4.85 pounds in seven or eight weeks, but we expect to get it down to six weeks with better birds. The feed company will pay you by the batch, by the month, or by the week. We like it by the batch, because you're less likely to spend it when you get a big check."

The Blands have to feed and water new chicks the first five or six days from pans they set on the floor by hand, but after that it is all automatic. They walk each house every day to keep the birds stirred up and to check the feed, water, heat, and ventilation to be sure everything is working. The temperature is especially important because the food goes to body heat, not meat, if the birds are too cold, and "a hot, humid day really melts 'em down," Forrest said. "We have a water fogger in the roof that puts out spray to cool them in summer.

"Each batch needs six inches to a foot of clean litter,

sixty bales of straw in each house. It used to take seven hours to spread the straw by hand, but now we can do it in 45 minutes with a brush hog. A brush hog? That's like a big, old, rotary power lawn mower mounted on the back of a tractor. That old brush hog really moves that straw around when it hits a bale. It takes a week after each batch of chickens to clean the houses, and we put the litter on our pastures.

"We own 171 acres and rent a hay field. All of our land is sowed down to grass. These old fields around here were all raw and eroded and red before we started putting chicken liter on them to build them up. I overdid it once. I lost 25 cows from the milking herd by feeding them hay that had four times the nitrogen they needed. We made *The Progressive Farmer* on that one! They ran a story as a warning to other farmers. We had used litter plus chemical fertilizers, and the hay was strong enough to kill a cow."

The Blands have switched out of the dairy business and are developing a beef herd out of the remnants of the dairy herd. They have 65 cows and one bull. They sell the old cows and breed the heifers. They sell the steers at a year and a half, when the price is right or when they feel like it. They plan to cut down because the cattle take too much time and they have to buy hay. "A few years ago we had a drought," Forrest said, "and we couldn't get decent hay. We had to pay $3 a bale for some old worthless stuff, so we began using chicken litter as a supplement. The cattle liked it so much we've been using it ever since. One-third grain and two-thirds litter is a fine cattle ration."

15

Patches and
Polka Dots

The South traditionally has been a one-crop farming area, but cotton did not hold undisputed sway over the entire South even in the days when it was king. Tobacco, peanuts, rice, and other specialty crops have had their own smaller fiefdoms, and now cotton itself has been relegated to the status of just another specialty crop. Each of the traditional specialty crops has been concentrated in its own particular niche within the South, and the areas of concentration have overlapped only slightly.

The specialty crops are cash crops. They have usually fetched good prices, and they have dominated the economies of the areas where they have been grown. The geographic concentrations of these crops originated by historic accident, when someone started planting the crop in an area that was environmentally suited to its production. A far larger area is suitable for growing them, but their acreage is limited by the amounts people can consume, which are fairly small. In 1982, for example, the United States had 64.8 million acres of soybeans and 9.8 million acres of cotton, but only 3.2 million acres of rice, 1.2 million acres of peanuts, 0.9 million acres of tobacco, and 0.7 million acres of sugarcane.

The specialty crop areas of the South are islands of cultivated land in a sea of dark pine forest. These crops are grown in small patches on the best land within these islands. They have rarely been grown on more than 5 percent of the total area even where they were most intensively cultivated. The islands of specialty crop production are scattered like polka dots across the map of the South.

The total volume of production of these crops has remained fairly stable for the last half-century because our consumption of these crops has not kept pace with our ability to produce them. Many people have stopped using tobacco and cut back on sugar for health reasons, Americans have never eaten much rice, and some Republicans, it is said, stopped munching peanuts when Jimmy Carter was president.

The general areas in which the specialty crops are produced have changed only slightly, but the margins of these areas have contracted because productivity has outstripped demand. Breakthroughs in technology doubled the yields per acre of these crops in the two decades after World War II. Farmers can produce the same amount on only half the acreage when yields are doubled (Figure 15–1), and the acreage of a crop must be reduced unless people start to use more of it. The combination of rising yields and slack demand has forced some marginal areas to reduce or even cease their production. Some land that once grew these crops has simply been abandoned because farmers have no acceptable alternative crop to grow on this land.

The broad outlines of the principal tobacco, peanut, rice, and sugarcane areas were formalized by agricultural legislation in the 1930s. Subsequent agricultural legislation has continued the allotment system of acreage controls and fossilized the geographic pattern of the 1930s. It has been extraordinarily difficult for an individual farmer to break into the system and start growing a crop if he does not have an allotment, and it is virtually impossible to start producing it in a virgin area.

The farmers who hold allotments have not been unaware of their good fortune, and they have treasured their allotments even when they have not been able to use them effectively. In 1985 the Federal Reserve Bank of Atlanta estimated that government price-support programs have added 40 cents to the price of a pound of tobacco, 28 cents to the price of a bushel of corn, 16 cents to the price of a pound of sugar, 11 cents to the price of a gallon of milk, 5 cents to the price of a pound of peanuts, 3 cents to the price of a pound of cotton, and 2.4 cents to the price of a pound of rice.

The farmers who own allotments have pushed for relaxation of the stringent rules governing their use, rental, and sale so that those with undersized allotments can sell or lease them to neighbors who want to expand their operations. A brisk business in buying, selling, and leasing allotments has developed as soon as the regulations have permitted such practices. The entire allotment system, and the geography of specialty crop-producing areas, could easily be transformed by the stroke of a pen in Washington, but it is guarded and protected by such strong and well-organized vested interests that changing it might be no easier than moving a cemetery or revising a curriculum.

Government programs have maintained the production of some crops in some areas even after it has become inefficient and should have been allowed to die. These moribund agricultural regions also have been maintained because they have the infrastructure necessary for the

Figure 15–1. Acres needed to produce units of specified crops. The number of acres needed to produce a given amount of a specific crop is the mirror image of the enormous increase in yields per acre since World War II. The combination of rising yields and slack demand has greatly reduced the acreage that is devoted to producing the traditional crops of the South. Reproduced by courtesy of the *Annals of the Association of American Geographers,* University of Minnesota.

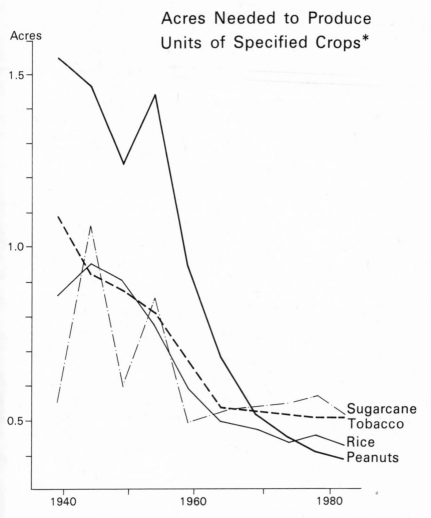

Acres Needed to Produce Units of Specified Crops*

Acres

1.5 —

1.0 —

0.5 —

Sugarcane
Tobacco
Rice
Peanuts

1940   1960   1980

*Peanuts 1,000 lbs.; tobacco 1,000 lbs.;
rice 2,000 lbs.; sugarcane 20 tons.

production of particular crops, while other areas have not been able to begin production of these crops because they lacked the necessary allotments and infrastructure. The South has been a splendid laboratory for studying the birth and death of agricultural regions.

The farmers in any agricultural region need specialized technical skills and knowledge. They must know how to distinguish boll weevils from June bugs, or know when and how to plant, transplant, top, sucker, and prime tobacco, or how to pull a levee in a rice field, or how to castrate a steer calf, or how to rig up a block and tackle to help a cow that is having trouble calving. They must understand fertilizers and pesticides. They must be good money managers, and they need bankers who will lend them the money they need to buy a new tractor, or a self-propelled combine, or a mechanical cotton picker, or a carload of feeder cattle. And they need to know how to repair the machinery when it breaks down.

A successful farmer also needs the right structures and facilities for producing, processing, and selling his goods: fences to separate crops and livestock; shelters for crops, livestock, tools, and machinery; on- and off-farm processing facilities such as cotton gins, tobacco barns, and packaging plants, to prepare the goods for market; and tobacco warehouses, grain elevators, stockyards, and auction markets where they can be sold.

16

Golden Leaf or Dirty Weed?

Tobacco was the first specialty crop that was grown in the United States. Virginia farmers started shipping tobacco to England eight years before the Pilgrims landed at Plymouth Rock, and they had already imported a boatload of slaves from Africa to work in their tobacco fields the year before the *Mayflower* arrived off the New England coast.

The tobacco plant has a chameleon-like ability to adapt to different environments. Only one single botanical species of tobacco is grown commercially in the United States, but each major tobacco district produces its own distinctive type. The leaf grown in Maryland south of Baltimore is light and chaffy and has little flavor, but it is prized by cigarette manufacturers because it burns so smoothly and evenly (Figure 16–1). The limestone soils of Kentucky and Tennessee produce a light leaf called Burley.

The leaf produced by early farmers on the fertile loam soils of central Virginia was strong and heavy, with a dark green color. Farmers cured it by hanging it in sheds to dry. Sometimes they tried to hasten the curing process by hanging the leaf over smoky fires. The final product was pretty harsh stuff by modern standards.

When farmers began to grow tobacco on the less fertile, light gray soils along the North Carolina line, however, the semistarved plant produced thin, mild, light green leaves. This leaf was so superior that counties began to boast about having the largest area of poor soil. Farmers learned that they could produce an even more desirable leaf if they continued semistarvation into the curing process, and they began baking the leaf over heated metal flues to dehydrate it rapidly.

The barns they used for flue-curing tobacco are squat

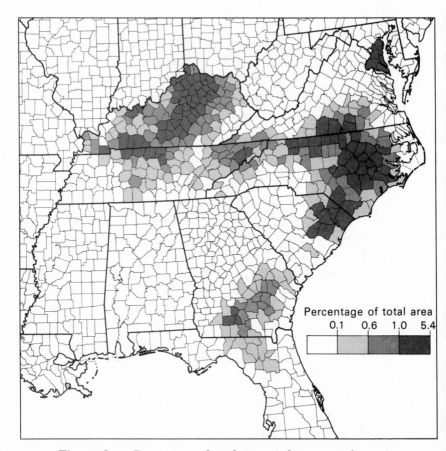

Figure 16–1. Percentage of total area used to grow tobacco in 1982. Bright or flue-cured tobacco is the principal variety produced in the Carolinas and Georgia, and Burley is the principal variety in Kentucky and Tennessee.

cubes, 16 to 20 feet on a side and 20 feet to the eaves. A lean-to roof at one end shelters a brick furnace, from which twelve-inch flues of sheet-metal pipe extend across the barn floor. The interior has a lattice of horizontal tier poles that are four feet apart horizontally and two to two and a half feet apart vertically. Farmers tied their green tobacco leaves to inch-square wooden tobacco sticks four and a half feet long and hung the loaded sticks on the tier poles. Then they fired up the furnace and gradually raised the temperature in the barn to about 180° F.

A complete cure normally took three and a half to five days, and changed the green leaf to a bright, orange-yellow color. This leaf has three names that are used interchangeably. It is called flue-cured tobacco because of the way it is cured. It is called Bright tobacco because of its color. And it is called Virginia tobacco in the export trade because it was first grown in Virginia, even though most of it is now grown in North Carolina (see Figure 16–1).

Flue-cured tobacco was used in the old days to wrap plugs of chewing tobacco because of its attractive bright color, but its main use now is in cigarettes, which did not really become popular until the boys came home from the trenches after World War I. The average cigarette contains 55 percent flue-cured tobacco, 33 percent Burley, 2 percent Maryland tobacco, and 10 percent oriental tobacco (imported from Turkey and Greece). The demand for flue-cured tobacco increased steadily when per capital consumption of cigarettes by Americans 18 years old and older rose from two a week in 1910 to one a day in 1918 to twelve a day in 1963, the peak year. It declined to less than ten a day in 1982 because more and more people have become concerned about the health hazards associated with cigarette smoking.

The production of flue-cured tobacco spread eastward and southward from the Piedmont of Virginia and North

Carolina at the expense of cotton (see Figure 16–1). Farmers on the Inner Coastal Plain of eastern North Carolina began to grow tobacco when the price of cotton dropped disastrously in the 1880s. Production spread into parts of eastern South Carolina in the 1890s and on to southern Georgia and northern Florida in the wake of boll weevil depredations in the 1910s.

Farmers on the Piedmont grew small patches of only two or three acres of flue-cured tobacco with family labor. They had log barns chinked with clay, and they used wood for fuel. In 1958 Jesse Tuttle, a tenant farmer near Germantown, North Carolina, told me that "people used to have wood choppin's in this country. We'd cut a while and talk a while and have a *big* time. I've went to 25 different places in one year, but that's all run out with these here chain saws.

"I was born and raised here," he said, "and I never done nothin' in my whole life but raise tobacco, nor did my Daddy. I'll tell you the truth, Greensboro [30 miles] and Mount Airy [31 miles] is the farthest away I ever been in my life. It takes thirteen months work a year to make a crop of tobacco. You've gotta get started on next year's crop afore you've sold this year's. Last year me and them ar wimmen made 'round four acres. I'm 68 years old, and my wife, she's 60, and my sister's 65, and my daughter, I reckon she's 'bout 38. She done finished school, and she can write an awful good hand. She drives that ar tractor. Next week I'm goin' to have my tobacco sprayed by one of these here airyplanes.

"We'll go out and prime half a day, and then string in the afternoon, and if we don't have enough to fill the barn we'll prime some more after we string. We always prime on Monday so's to be through on Saturday afternoon. I cured on Sunday last year, but I shouldn't ought to had, 'cause the Good Book says to keep that day holy. Man grows tobacco all his life and happens to miss Heaven, he's done lost it all. . . ."

The hard-working, God-fearing, native white farmers of the Piedmont, who accepted the fundamentalist religious doctrine that hard work and suffering on earth would result in a better life hereafter, were able to produce flue-cured tobacco with a much smaller labor force than farmers in other districts.

The farmers of eastern North Carolina retained their traditional tenancy system when they switched from cotton to tobacco, and they grew the crop on a large scale with black sharecroppers. They used wooden frame rather than log for their barns, and sheathed them in black or green tarpaper held in place by vertical wooden battens nailed to the wood beneath. They needed several barns because one could only house the tobacco from three to four acres. They grouped the barns in clusters for convenience, but far enough apart to keep all from going up in smoke if one caught fire, as often happened.

The flue-cured tobacco farmers of eastern North Carolina needed many workers. The labor required to grow one acre of tobacco in the 1950s could have produced 7 acres of cotton, 35 acres of corn, or 90 acres of soybeans. Transplanting required 20 hours an acre; topping and suckering, 30; harvesting and curing, 350; and preparing for market, 50. Unlike other farmers, whose labor needs were reduced by mechanization, the labor requirements of flue-cured tobacco farmers actually increased when their yields per acre increased because they needed more hands to pick the more abundant crop.

Tobacco must be started in special seedbeds. The farmer planted far more seedbed than he needed, and at transplanting time in the spring he selected only the best and most uniform seedlings to ensure a uniform crop that would be easy to harvest. Transplanting normally took a crew of four or five people. Farmers liked to transplant at the dark of the moon. Some people though they were just being superstitious, but the light of a full moon on a clear spring night with crisp temperatures brought out hordes of cutworms, one of more than a hundred insects and diseases that enjoy tobacco almost as much as people do. Tobacco farmers were constantly battling pests, and they tried to minimize their ravages by growing the crop no more than once every three or four years on the same ground. During the growing season children too young to do anything else were given big buckets and sent into the fields to pull worms off the tobacco plants. When the bucket was full they threw the worms into a fire and started all over again.

Harvest began at the end of July and lasted for six to eight weeks. The leaves ripen from the bottom up at a rate of about one every other day, and the farmer tried to prime, or pick, them as soon as they became ripe. He harvested the field, filled the barn, and cured the crop once a week. The usual harvest crew for three acres of tobacco, enough to fill a barn, consisted of two primers, a sled driver, two stringers, four handers, and two helpers, plus innumerable, squalling small-fry and yelping hound dogs. Even farmers who were below the poverty level usually had to hire help for the tobacco harvest.

A pair of primers, stooping from the waist, tore the two to four ripe bottom leaves from each plant, and placed them in a mule-drawn sled with waist-high burlap sides. A child drove the sled back to the barn when it was loaded, and teams of two handers and a stringer tied the leaves to tobacco sticks. A hander picked up several leaves, arranged them into a neat bunch, or hand, and passed them to the stringer, who tied each hand to a tobacco stick. The loaded sticks were hung on the tier poles in the barn. They were taken down when they had been cured, and a new batch was hung. The leaves had to be untied from the sticks and sorted into grades before they could be taken to the sales warehouse.

A tobacco sales warehouse is a vast, cavernous structure. Rows of skylights on the low roof do little to pierce the gloom on the open sales floor, which is larger than several football fields. Piles of tobacco are lined up in rows on the barn floor and sold by auction. The chanting auctioneer walks slowly down one side of a row facing the team of buyers on the other side. Each warehouse has a team of six buyers, one from each of the cigarette-manufacturing companies. The United States has only six cigarette companies, but they make and sell more than 170 different brands. The buyers bid with hand signals because the auction is much too fast for voice bids. The auctioneer sells a pile of tobacco once every eight to ten seconds.

Tobacco resisted mechanization longer than any other major field crop produced in the United States. As late as 1960 some observers doubted that machines could ever perform some of the delicate tasks required in growing the crop, and its mechanization still is far from complete. The labor needed for an acre of flue-cured tobacco dropped from 425 hours in 1965 to "only" 175 hours in 1979, but as late as 1984 an acre of Burley tobacco, which is produced with hand labor on small farms in Kentucky, Tennessee, and the mountains of western North Carolina, still required 300 to 350 hours of hard work.

The various phases of tobacco production have been mechanized quite unevenly, and mechanizing one operation has often turned another into a bottleneck. The full potential of mechanization can be realized only when the total technology has been adopted. Mule- or tractor-drawn transplanters replaced transplanting by hand before World War II, but they still require a driver and two workers for each row. Chemical sprays to control suckers came onto the market in the late 1950s. They made life much easier for tobacco farmers, but they did nothing to allay the fears of those who were already concerned about the health hazards of tobacco. Topping machines came into widespread use in the late 1960s. They slice off the flowering seedheads with large rotating blades, like the blades of a power lawn mower.

Harvesting has been the hardest part of flue-cured tobacco production to mechanize because the identification of ripe leaves requires human judgment. The first harvesting machines were called taxi rigs, because primers rode on them instead of having to work bent over all day. Taxi rigs are cumbersome contraptions with bucket seats for four primers near the ground and a platform eight or nine feet above it. The seated primers pull the ripe leaves from the stalks as the machine wobbles across the field. They attach the leaves to a conveyor belt that carries them to the platform above, where other workers tie them to tobacco sticks. Taxi rigs merely modified labor requirements instead of reducing them. Women, children, and men not sturdy enough for the backbreaking labor of priming on foot could be employed on taxi rigs.

Fully mechanized harvesters came onto the market in the late 1960s. They reduced labor needs appreciably because one person can operate them. The operator sets blades that cut off all the leaves between specified heights on each stalk. Some leaves in this height zone are already overripe, and some are still green, but the sacrifice in quality is repaid by the enormous saving in labor.

The development of mechanical harvesters was associated with the replacement of flue-curing by bulk-curing. The product is still called flue-cured tobacco, but it is cured by forcing hot air through it instead of depending on convection from a flue on the barn floor. Workers clamp loose tobacco leaves in metal racks and pack them tightly in the new, metal bulk-curing barns, which have heating systems with precise temperature controls. The contemporary countryside of flue-cured tobacco districts is an odd blend of the past and the future. Tens of thousands of tired, old flue-cured barns, squat cubes with sagging metal roofs, have been replaced by batteries of sleek, new bulk barns that look like semitrailers waiting patiently to be attached to a truck that never seems to arrive.

The adoption of a new technology almost inevitably forces farmers to enlarge their operations because the real cost of a piece of farm equipment is closely related to the amount of ground on which it is used. A farmer cannot afford a new machine unless he has enough land to keep it busy. In 1982 the average Burley tobacco farmer raised only three acres of the crop because Burley still had not been mechanized, but the average on flue-cured tobacco farms increased from eleven acres in 1972 to twenty acres in 1982.

More than four-fifths of all flue-cured tobacco farmers have expanded their operations by renting tobacco rights from other farmers. Dispersing their production onto scattered parcels of rented land reduces the chance that the entire crop will be drowned by torrential showers that overload field drains, riddled by hail, or parched by the failure of thunderstorms to hit a particular field.

In recent years tobacco farmers have had to cope with a complicated and controversial government price-support program. The original allotments were based on the acreage a farmer grew in 1938. The guaranteed price encouraged many farmers to produce and sell as much tobacco as they could, regardless of its quality, and in 1965 the basis for the program was changed from acreage allotments to poundage quotas.

A poundage quota, in effect, is a license to sell a given weight of tobacco. A tobacco farmer leaves part of his crop standing in the field if his allotted acreage produces more pounds than he is permitted to sell because harvesting and marketing account for three-quarters of his total production costs. In a good year travelers may be surprised by the amount of tobacco that remains in the fields unharvested.

The principal beneficiaries of tobacco programs have been the original recipients of allotments and quotas, who paid nothing for rights that have become quite valuable. Young farmers have criticized the programs because they add to the cost of getting started. People who have bought farms with tobacco rights defend the programs because they would lose their investments if the programs were abolished.

In 1985 quotas rented for 50 to 70 cents a pound, or $1,000 to $1,500 an acre. They could be leased or sold only within a county. A study of geographic variations in lease rates in 1968 concluded that 123 million pounds of tobacco would be transferred from the Piedmont to the Coastal Plain in North Carolina if quotas could be moved across county lines, and 67 million pounds would be transferred from the Piedmont of Virginia and North Carolina to the Inner Coastal Plain of South Carolina and Georgia if quotas could be moved across state lines.

Flue-cured tobacco farmers have lost part of their traditional market because the government support price for the poorer grades has been too high. This price umbrella has enabled producers in Brazil, Zimbabwe, and other countries to take over part of the world export market and even to make inroads in the American domestic market. In 1982 a third of the flue-cured tobacco used in American cigarette factories was imported. The use of domestic flue-cured tobacco has also been cut back by reductions in the circumference of cigarettes and by the switch to filter cigarettes. The amount of tobacco used to make a thousand cigarettes dropped from 2.7 pounds in 1950 to 1.7

pounds in 1980, and cigarette manufacturers started using harsher grades of tobacco to counteract the flavor-sapping effect of filters. Farmers get only six or seven cents for the tobacco in a pack of twenty cigarettes, but in 1982 the Federal government collected $4.6 billion in cigarette taxes, and state governments collected another $4.1 billion.

Tobacco farmers have been under fire from those who dislike smoking, and critics have argued that they should shift to some other form of farm enterprise. Many tobacco farmers would be happy to oblige, but they have no acceptable alternatives. Their farms are too small for livestock operations, and no other crop could produce the same return per acre as tobacco. Vegetables are often suggested, but vegetables are risky, and they require a sophisticated and expensive infrastructure for marketing. The tobacco-producing districts are among the few remaining islands of agriculture in the South. Terminating tobacco production would drive yet another nail in the coffin of southern agriculture, and it would cause great suffering for tobacco farmers.

Noah Hardee, 57, raises flue-cured tobacco east of Greenville, North Carolina, in the heart of the nation's leading flue-cured tobacco district. The land is flat, sandy, and poorly drained. The grayish brown soil contains few plant nutrients, and farmers must fertilize it heavily in order to grow crops. Long, straight drainage ditches overgrown with weeds border most fields.

One nice day in 1982 Noah and Dave, one of his hired men, were burning plastic chemical containers in a rusting 55-gallon drum. "I'd like to spray today," Noah said, "but I can't because the wind is blowing at least ten miles an hour. You've got to be awfully careful with chemicals. The first poison we used was arsenate of lead and Paris green, and it's a wonder that I'm still alive to tell about it.

We dusted it on with a machine behind a mule, and by the end of the day we were as white as that house over there."

Noah's father started off with cotton as his main crop, but he switched to tobacco around 1900. He had 90 acres of tobacco in 1933, but the government program cut him back to 60. He had eight croppers, four white and four black, plus four hired laborers. Noah worked for his father before he went into the service in World War II.

"I was raised up as a tobacco farmer," he said, "and when I got home from service that was all I knew. In 1947 I started farming with 8.5 acres of tobacco, 15 acres of wheat, and 12 acres of corn. I grew wheat as a cash crop and corn for hogs and chickens. I grew 6 acres of cotton until 1964, and I also grew peanuts, but I forget the area. I quit cotton because I couldn't get labor, and I quit peanuts because my allotment was too small. I sold it to a fellow that was buying up peanut allotments to enlarge his operation.

"We used to say that 35 acres was about the right size for a farm. I inherited 32 acres. I added 35 more in 1952 because I'd bought a tractor and other equipment, and I've been adding little bits ever since. I own 100 acres, and lease 300–350 more in eleven different farms, all small and scattered. I lease it because I can't afford to buy it. I used to have an 18-acre tobacco allotment, and I hired 12 or 13 people to work it. Now my allotment is down to 6 acres. Two young men are my only full-time labor, but I work like a hired man myself. The farms I rent have a tobacco allotment of 36 acres."

Noah plants his tobacco on his best land because his allotment is so small. He chooses well-drained soil, even though it is fairly sandy. He said the longer you can stay off a piece of land with tobacco the better crop you get. He plants his seedbed at the end of January, and transplants in mid-April with a four-row mechanical transplanter, which needs nine workers.

He sprays to control suckers and insects, and he uses machines to top and harvest. He used to pick by hand five or six times, but now the machine only picks two or three times. He is trading quality for labor, and he would much prefer to prime by hand if he could get labor. He has fourteen conventional flue-cured barns, but not a one of them is usable. He switched to bulk barns in 1970 to save labor and fuel. He owns six and leases two more. His yields run around 2,400 pounds per acre.

He used to sell his tobacco in Georgia. There are no processing plants in Georgia, so the empty trucks could haul a load of leaf back cheaply after they had delivered leaf from Georgia to processing plants in North Carolina. The market opens earlier in Georgia, and foreign buyers thought that the Georgia leaf was better, so they paid a higher price for it. It could be sold loose in Georgia, but government regulations required Carolina farmers to bundle it into hands for sale until the mid-1960s. The government program now requires the farmer to designate one warehouse and sell it all there, and the warehouse manager tells him how many pounds he can sell each week.

The rest of Noah's land is in 145 acres of corn, 120 acres of double-crop wheat and soybeans, and 60 acres of full-season soybeans. He averages 100–120 bushels of corn per acre, 38–42 bushels of beans, and 45 bushels of wheat, and he sells them all. "Crops are a one-time-a-year thing," he said. "They only give you income once a year, and livestock give me a good cash flow. I went into livestock to diversify. In the early 1950s I built a small brooder house. You could sell eggs for eighty to ninety cents a dozen and they paid the bills. I worked up to 2,500 laying hens, but the market got so poor that I got out. I was just a producer, and I needed someone to market for me. In the early 1960s I went into hogs. After three or four years I built a farrowing house, but some people bought the property back of me and started a housing development with a lake. I looked down the road a ways and knew I was headed for trouble because my land drains into that lake, and they were right next door to my hogs, so I backed out. I had been growing into hogs, and I just grew back out."

In 1968 he went to broilers with Frank Perdue. He started with two houses that were already outdated by the time they were completed, and now has four that hold 15,400 birds each. One man works full-time in the poultry houses. Noah tries for five and a half flocks a year.

"I am happy to have the income from broilers and from the other crops," he said, "but when you come right down to it basically I'm a tobacco farmer, like the other farmers in this part of east Carolina. Tobacco is our life blood. I don't know what we'd do if they tried to make us stop growing tobacco because no other crop can produce nearly the same return per acre. It would force us to stop farming."

17

Peanuts and
Pastures

In 1919 the good citizens of Enterprise, Alabama, erected a monument to the boll weevil because it had brought them new prosperity when it forced local farmers to switch from cotton to peanuts. "The world's only monument to a pest" graces a fountain at the town's principal intersection.

The area east of Enterprise along the Alabama-Georgia line has become the nation's leading peanut-producing district (Figure 17–1). This area anchors the southwestern end of the Inner Coastal Plain, the second most important agricultural area in the South. The northeastern end of the Inner Coastal Plain is anchored by the other major peanut district, along the Virginia-Carolina line. Cotton was grown on the Inner Coastal Plain, but the sandy loam soils are a bit droughty for cotton. The level topography is well suited to large fields and large machines, however, and this area really came into its own as an agricultural region after World War II.

Its relatively late start handicapped the agricultural development of the Inner Coastal Plain because government programs closely controlled most of the crops for which it was best suited. Corn has been the principal food and feed crop, but the soils are so droughty that yields are poor unless the land is irrigated. Cotton is still hanging on in a few isolated pockets, but the cotton acreage is only a fraction of what it once was. Soybeans have been the salvation crop. They have not been restricted by government programs, and their price has been fairly good. Their acreage has increased enormously, and they have become the leading crop in most of the region. The exceptions are the fortunate areas that were already growing tobacco or peanuts when acreage allotments were imposed.

Peanuts were brought to the United States from Africa on slave ships. They were grown widely in the South before the Civil War. Yankee soldiers developed a taste for them during the Virginia campaigns, and the Virginia-Carolina area began to specialize in peanut production to satisfy the new national demand. Farmers in the Georgia-Alabama area turned to peanuts after the boll weevil wiped out their cotton crops around the time of World War I. Farmers were urged to grow more peanuts during

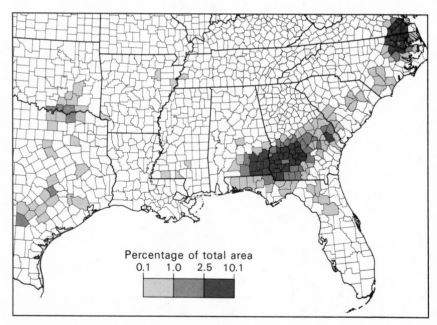

Percentage of total area
0.1   1.0   2.5   10.1

Figure 17–1. Percentage of total area used to grow peanuts in 1982.

both world wars because they are a good source of vegetable oil.

Peanuts are not nuts, despite their name. They are legumes, close relatives of beans and peas. They grow on bushy plants only a foot or so tall. The stems of the plant droop to the ground after its flowers have bloomed, and the nuts mature in shells beneath the soil, but not on the roots of the plant. The nuts are a high-protein substitute for meat, and the vines were made into hay for mules. The peanut areas of the South were also major hog-raising areas. Farmers planted alternating rows of corn and peanuts and turned hogs into the field when the crops were ripe. Some people say that a real Virginia ham can come only from a hog that has been fattened on peanuts.

At harvest time farmers dug up the plants, shook the dirt from the nuts, let the vines wilt on the ground for a few hours, and then stacked them to dry on poles eight feet high. The slender stacks, no more than three feet thick, stood in the fields for four to six weeks until the pods were dry enough to be threshed from the vines. The harvest has been mechanized since World War II. Special plows dig, shake, and windrow the vines. The vines are allowed to dry for a few days, and then the pods are removed by combines that deliver them to metal wagons in which they are dried by hot air.

The average American consumes seven pounds of peanuts and peanut products each year. Half is peanut butter, a quarter is salted nuts, and a quarter is in candy. Most of the peanuts produced in the Virginia-Carolina area are salted for direct consumption, but much of the Georgia-Alabama crop is processed. In other parts of the world peanuts are a major source of edible oil for human consumption, and the protein-rich meal is fed to livestock, but in the United States peanut oil and peanut meal have to compete with soybeans, cottonseed, and sunflowers.

Government acreage allotments and marketing quotas have been in effect for peanuts since 1949. A new program was started in 1982 because peanut yields had increased from 1,000 pounds per acre in the mid-1950s to more than 2,600 pounds per acre in the late 1970s, and peanut farmers were producing more than the market could consume. A two-price poundage quota program was started in 1977. A complex formula determined the poundage each farmer was permitted to sell from his allotted acreage. The peanuts he produced in excess of this quota were called additionals. The price for quota peanuts is set by law, and the price for additionals is based on the world market price. Since 1981 poundage quotas have been reduced regularly in an attempt to reduce surplus production and to put quotas in the hands of actual producers.

In 1982 Jack Fleet grew 240 acres of peanuts on a farm six miles southeast of Colquitt, Georgia, in the center of the Georgia-Alabama peanut area. "I have an allotment of 375 acres," he told me, "but I cut down on my acreage when they changed from acreage to poundage. The government put us on quotas when we popped the yield from 1,000 to 4,000 pounds. My quota is about 50 percent of my normal yield. I plant about 60 percent of my allotment. That gives me enough to make my quota, plus a little bit of room to stumble.

"The support price for quota peanuts is $550 a ton," he said, "but the support price for additionals is only $180, which is less than the cost of production. The production cost is influenced by land rent, allotment rent, irrigation rent, type of irrigation, land quality, and so many other variables that I can't give you a simple figure, but on this place I would guess that it's probably around $600 an acre."

Jack Fleet was born in Warsaw, Missouri, in 1923 and has

a degree in agriculture from the University of Missouri. He wanted to farm, but could not get land, so he went to work for Doane Agricultural Services. "I was in this area working for Doane," he said, "and met an old boy over in the next county that had a lot of land to rent. All I had was $10,000 and four kids, but he backed me to farm on shares, and we moved here in 1964. I worked for him for a while, but we fell out, and I leased this place, Babcock Plantation, in 1969. I bought a half interest in the brood cows and have upgraded them, and I talked the owners into starting to irrigate."

Jack has a lease contract, fifty/fifty on the livestock and crops, with Babcock Enterprises of Pittsburgh, Pennsylvania (Figure 17–2). They have owned this place since the 1890s, when Mr. Babcock bought a huge acreage for timber. It was all virgin pine then. Jack's house is on the old sawmill site, and narrow-gauge railroads spoked out from it. The lake in back of his house is the old log pond. Babcock had a regular town here, but nothing is left of it now except a few foundations and some pilings in the pond that are visible at low water.

They had cut out all the pine by the 1920s, so they sold off the land for whatever they could get for it. They still had 20,000 acres after World War II and cut off the second-growth pine, but Mr. Babcock's son, Fred, wanted to farm, and he ordered them to stop selling the land. He started cattle and had a big feed lot, but he never did make any money on it, and they finally closed it down.

Local farmers used to grow cotton, but it finally disappeared in 1972 or 1973. The boll weevil and the boll worm were terrible, and many black people went north. Outlawing DDT finished it off because farmers had to poison with it 28 to 30 times a year. Corn, not peanuts, has taken the place of cotton, although peanuts have been the money crop in the area since World War II. Jack said that once they were grown by sharecroppers, but that was long before his time.

He plants peanuts with an airplane around April fifteenth, after the bad windy season is over. He also uses an airplane to spray them with fungicides every two weeks

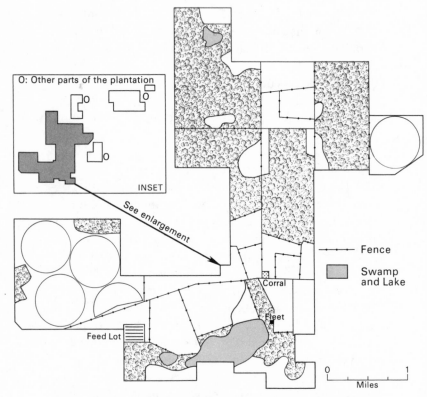

Figure 17–2. The Babcock plantation. The Babcock plantation consists of five separate parcels, but the enlarged map shows only the largest. The cropland is watered by five center-pivot sprinkler irrigation systems. Much of the unirrigated land is wooded, and the rest is open pasture.

and to spray insecticides when he needs them. He harvests peanuts in September and October with a digger inverter that turns the nuts up. He lets them cure on the ground for two or three days, and then combines them with peanut pickers that cannot harvest anything but peanuts. He hauls the peanuts back to the shed in dryer wagons, hooks the wagon up to the dryer, and dries them with propane gas. The dryer shed holds 10 wagons, and he needs 20 to keep it full. He dries the peanuts overnight, and then hauls them to the mill in Colquitt to be shelled. His yield has gone from 1,000 to 4,000 pounds per acre because of new varieties, good new fungicides, and irrigation.

The Tifton sandy loam soil is his best peanut land, but the owner wants to keep part of it in trees for environmental reasons. The Troup sand soil was marginal to submarginal before irrigation and chemical fertilizers, and all it grew was poor pines. It really was not even good for pines. Jack has been building it up by adding stalks and other crop residues, and every year it gets better. It is good peanut land when it has been built up.

"All of this soil is so sandy that we're only a week away from drought," he said. "We need an inch and a quarter of rain every Saturday during the growing season. Irrigation has made a tremendous difference. I have eight center-pivot sprinkler irrigation systems that can irrigate more than 1,000 acres. We've got an abundant supply of underground water, and anything we withdraw is quickly replenished by torrential summer and fall rainstorms.

"We have a terrible blowing problem on this sandy land, and we have to protect the soil with a winter cover crop on all the cropland. We disc right behind the peanut pickers and sow the cover crop, mostly rye, but some wheat. We combine some of the rye for seed for the next crop, but a good portion we just plow under because it makes a good fertilizer. This sandy land leaches pretty badly, but the cover crop ties up the fertility until we can plant another crop."

Jack grazes his cattle on the cover crop in winter, then plows it under before he plants corn in the spring. He likes to have a three-year rotation: a year of peanuts and then two years of what he calls a grass crop, either corn or grain sorghum. Peanuts do not respond too well to direct fertilization, but they are good users of residual fertility after a heavily fertilized crop like corn, and corn helps to control diseases in peanuts.

He plants corn in February or March and starts to shell it around July fifteenth. He discs the land immediately, then plants forage sorghum to control weeds because "those weeds just come a-boilin' up after you combine corn." He averages about 125 to 140 bushels an acre, and he tries to sell the entire crop before the Corn Belt crop comes in. There is always a big price break when the Corn Belt crop hits the market, but most of the year he has a price advantage in corn because the South is a corn-deficit area.

Jack Fleet has about 1,000 acres of land in the peanuts-and-corn rotation. Peanuts always get the best land, the irrigated land, and he puts other crops on the rest. He has 400 acres of soybeans and gets about 30 bushels to the acre. They are a stepchild crop that he grows on dry land and on odd-shaped fields that he cannot irrigate. They fit the same place in the sequence as peanuts. The soil needs a legume, and it is good to change crops. They are a second-best crop, but they can be a good money crop. They are easy to produce, and he can produce them at half the cost of corn. He double-crops all of his beans with small grain, either wheat or rye. He combines the beans in November, sows grain in December, starts to combine the grain about May fifteenth, and plants beans in the stubble.

The cattle business ties the whole operation together.

The Babcock Plantation has 6,000 acres, but only about 2,000 acres is cropland, and most of the cropland is in forage crops that can only be used by cattle. More than 1,000 acres of the plantation is improved pasture that he seeds and fertilizes in January after he burns off the old dead grass, and another 1,000 acres "is the kind of land that just sort of holds the world together, piney woods that is poor pasture and poor timber."

Jack runs cattle on the winter cover crops and on the marginal land that will not grow good crops. Cattle use labor in winter that would otherwise go unused, and they help his cash flow. He said that a lot of people just use cattle to fill in and do not make much money with them, but he believes in managing them intensively, and he will not keep a cow unless she is really producing. He told me that the brood cows that were on the place when he rented it were a pretty sorry lot of Angus, all short and dumpy. He has upgraded them with Brahma or Charolais bulls and has 700 first-cross brood cows. They calve from January to March. He weans the calves and sells them the next spring right on the farm to buyers who come from all over the South.

Jack and his wife have incorporated their operation for inheritance-tax purposes. Half the stock is in her name and half is in his. His son, who has a degree in agronomy, works on the farm for a salary and a share of the profits. Jack also has a cowboy, a field manager, and one worker "who is just there." He has seven tractors, a small bulldozer to clear land and pull tractors out of boggy spots, a combine, two peanut pickers, twenty peanut trailers, a ten-trailer dryer, three big grain bins, and six smaller ones. He estimates that since 1969 he has built up an estate of 1.5 million dollars in cattle, machinery, and equipment.

"The cattle operation is necessary to tie the whole thing together," he said, "but when you come right down to it this is really peanut country. Most of the peanuts in the United States would be grown within 100 miles of here if the support program was abolished, but the support program has kept other areas in peanuts. Soil and climate are the main reasons for peanuts in this area, but the shellers are here, and the farmers have the skills and equipment needed to handle the crop. We've got the best peanut growers in the world."

18

Rice for the Rest
of the World

We're going to spray tomorrow morning,"
said Tommy Wollams. "Be here at ten o'clock if you want
to watch the cropdusters." Two big, red-and-white tank
trucks had already arrived at Tommy's rice field when I
got there, and they were parked beside a dirt landing
strip. Along the way they had pumped water out of the
swamp to use in mixing the chemicals. Shortly thereafter
we were joined by a pickup truck towing a large tank of
propanyl and loaded with green-and-yellow boxes of Basa-
gran and drums of Bolero.

Then we heard the planes, sleek single-engine mono-
planes with 50 spray nozzles on the trailing edges of their
wings. They landed in great clouds of dust and taxied to
the trucks. Bob and Will climbed down from their bubble
cockpits. They took off their helmets, and Bob covered his
bald head with a brown baseball cap. He is growing gray at
the temples. "There are bold pilots," he said, "and there
are old pilots, but there are no old, bold pilots." Will is
younger, with dark tousled hair, muttonchop sideburns,
and a fringe of beard along the edge of his jaw. He seems
to be less concerned about longevity.

The truck drivers began mixing the chemicals and filling the tanks in the wings of the planes with a six-inch
rubber hose while we discussed the day's operation. On
the wing of one of the planes Tommy spread out a map of
the field that was to be sprayed. Bob punched his pocket
calculator and figured that it would be most efficient to

cover the field in swaths 39 feet wide. "What's it going to cost me?" Tommy asked. Bob did some more calculating and figured $36 an acre for the chemicals plus $3 an acre for the planes: $27,300 for spraying 700 acres one time.

"Why do you have two planes?" I asked.

"Farmers like planes to work in pairs because they're used to it," Bob answered, "and it keeps the flagmen busy. One plane is reloading while the other is spraying. The two have got to fly at the same height above the ground, or you'll get uneven coverage."

They took off when they were loaded, and Tommy and I drove to the edge of the field to watch them. Great rolls of spray eddied behind them as they buzzed across the field six feet above the ground. At the edge of the field they turned off the nozzles, banked steeply above our heads, and flew back to make another pass. Flagmen were stationed at the opposite ends of the field to show them where to fly. After each pass the flagmen took thirteen long strides to the left and waved their flags to mark the next pass. They were doused with spray on each pass, and Tommy remarked that they both had better take good long showers after the job was finished.

Tommy Wollams, 30, is a tough old boy. He grows seed rice, which must be especially clean, at the western end of the Texas rice belt, about a 100 miles west of Houston. His farm is surrounded by post oak woods north of Edna, Texas, and is well removed from other rice farms to protect the crop against contamination. When I was in Texas in 1982 he volunteered to meet me at the Morales store. There *is* nothing else in Morales.

We got into his battered, blue pickup truck and drove fives miles north on the blacktop, then turned east on an unpaved road. "This old washboard road costs me $5,000 a year," he said, as we banged along. I didn't think he was getting his money's worth, but he explained, "I tear along it too fast when I need a part for a machine that has broken down at a bad time and that really runs up my bills for front-end alignment.

"I'm just a poor boy farmer," Tommy continued, "but I do know how to grow a good crop of rice. I rent this land from Mr. Tom D. Henderson, who produces and sells seed rice. He has a big rice drying and storage operation over in El Campo. Twenty-five years ago he flew his own plane all over looking for good land to buy. He bought 2,100 acres of land in this patch. It was all pretty much worthless black-jack oak or post oak, they're the same thing, but there was also some live oak on it. He cleared it all himself. Two-thirds of it lays out each year as pasture, and he rents it to cattlemen. He only has enough water for 700 acres of rice, and he can afford to let it lay out. He is really rotating a year of rice with two years of pasture by renting it to different people."

Tommy leases 700 acres in a share operation. The owner provides the land and water for irrigation, Tommy provides the equipment and the labor, and they split the income fifty/fifty. Tommy has about $150,000 worth of equipment. He still owes money on it, and he leases some of it. He has six tractors—two 175 horsepower, one 155 horsepower, one 135 horsepower, a 125 horsepower, and a 70 horsepower—two combines and "all the usual implements—breaking tools, planting tools, different kinds of plows, all that sort of thing." He employs one full-time Mexican worker, who lives in a trailer that is the only building on the farm, and one part-time worker. The farm has five water wells, but none of them is very good. The wells feed the main canal that runs through the center of the farm, and laterals from it feed the units on either side. Tommy has divided the farm into 100-acre units for water management.

"It's safe to plant rice when the buds on a pecan tree are

as big as a squirrel's ear," Tommy said. "That old pecan tree won't lie to you. I found a big old one back in the swamp, and it tells me when it's safe to plant, usually late in March. Then I flood. You have to check the pumps twice a day when you're flooding to be sure they're working. We combine the rice in September. We unload the grain from the combines into auger wagons in the fields that haul it to trucks on the road. The trucks would bog down if we took them into the fields. Last year I had 5,500 pounds per acre.

"This land is right at the northwestern edge of the rice belt. The soil is light and sandy, not like the black soil farther south. There's six inches of sand over a yellow clay subsoil that holds the water pretty good. You can't grow row crops on this sandy soil. It's too hot and dry, and they just burn up. The only reason we can grow rice here is because we can irrigate it. I am about as purely chemical as you can get. That sand just holds the roots of the plant so I can feed it."

Rice is yet another specialty crop that is grown in islands in the South (Figure 18–1), but, unlike tobacco and peanuts, it is not grown on land that once grew cotton because it needs a different kind of environment. Two-thirds of the American rice crop is exported. Half the world's people live on rice, but Americans do not like it very much. In 1982 the average American consumed 114 pounds of wheat flour, 65 pounds of potatoes, and only 12 pounds of rice. We consumed a quarter of our rice in the form of beer. We ate about 60 percent directly, and the rest was processed into soups, cereals, and baby food.

Half the world's rice is consumed within five miles of the paddy where it was grown, much of it by the very people who grew it. The United States produces only 2 percent of the world's rice, but this country supplies 20 percent of the rice that enters world trade. The world rice market is unstable because it is subject to considerable manipulation

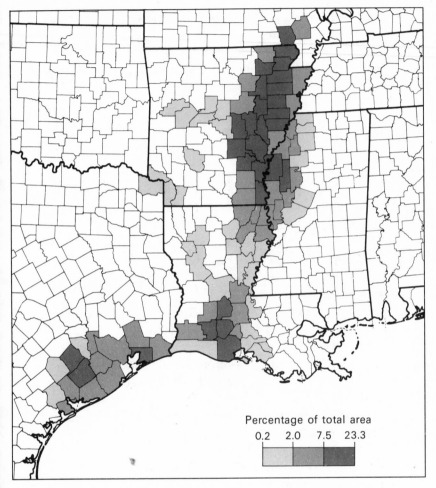

Percentage of total area

0.2   2.0   7.5   23.3

Figure 18–1. Percentage of total area used to grow rice in 1982. The principal rice-producing areas in the South are the Delta of Arkansas and the Coastal Prairies of Texas and Louisiana.

by governments whose citizens depend on rice, and the rice-eating countries of the world are not noted for their affluence. American rice exports have been subsidized by such devices as the Public Law 480 (Food for Peace) program, which permits low-cost sales or outright gifts of food, including rice, to poor countries.

The rice plant is a native of tropical marshlands, and it likes a hot, moist environment. It needs high temperatures during the growing season and abundant fresh water for irrigation. The crop needs land that is flat enough to be irrigated, but has a slope of two to three feet to the mile so it can be drained fairly quickly. The subsoil must be impermeable enough to avoid excessive water loss by seepage.

The rice farmer begins his crop year by leveling off his fields with a giant scraper called a land plane. After he has smoothed a field he divides it into "cuts" by using a special tractor-drawn plow to "pull" levees 50 to 100 feet apart that snake along the contours. The contour interval between levees is 0.2 feet on level land and 0.3 feet on "steep-sloping" land. The levees are 18 inches high to hold water in the cut, and 9 to 10 feet wide to let the farmer drive his machinery across them easily. Each levee has a staggered line of metal or plastic gates, 3 to 9 feet wide, that can be opened to fill the next lower cut when one cut has been flooded.

The farmer may plant rice seed either on dry ground or in a flooded field. A dry field must be flushed (flooded and drained immediately) after it has been planted unless there is enough rain to keep the soil moist. The levees obviously must be pulled before an airplane can plant seed in a flooded field. A wet-seeded field must be drained to parch aquatic weeds as soon as the rice plants stick their heads above water.

A rice farmer normally floods his fields two to four weeks after seeding, when the plants are six to eight inches high, and he maintains a water depth of three to six inches for 60 to 90 days, until two or three weeks before harvest. Flooding a field helps to minimize loss of nitrogen and to control weeds, although a farmer might have to drain it temporarily in order to get into it with ground machinery if it becomes infested with diseases or insects.

Rice needs large amounts of nitrogen at just the right time in order to produce top yields. The rice farmer must apply nitrogen fertilizer before the plants are 20 days old and again just before the grain begins to form at the top of the stalk. He is also a heavy user of other agrichemicals, such as herbicides to kill weeds, pesticides to kill insects, and fungicides to control diseases. Many of these chemicals are applied by cropdusters.

Every acre of rice grown in the United States is irrigated. The rice farmer may pump his irrigation water from wells or streams, or he may buy it from a canal company. Water straight from a well may be so cold that it harms the plants, which become uncomfortable when the water temperature gets below 75° F or above 85° F. Some rice farmers have constructed reservoirs where they can give the water a chance to warm up. Some farmers still use open ditches to carry water from the reservoir to the fields, but underground pipes are better. Ditches become overgrown with weeds, they are expensive to clean and maintain, and they are easily damaged by burrowing animals.

Rice farmers drain their fields and harvest their crops with self-propelled combines when the stems start to droop under the weight of the seedheads. At harvest the grain contains 18 to 22 percent moisture, and the farmer must reduce the moisture content to 12 percent as soon as possible to prevent spoilage. Many farmers have their own grain-drying facilities, but some still haul their rice to a commercial drier to be dried and stored. Enormous batteries of elevators for storing rice dominate the small towns in rice areas.

Immediately after he has harvested his crop the rice farmer discs the stubble to prepare the field for the following crop. Most farmers rotate rice with two years of soybeans, which makes it easier for them to control weeds and to prepare the field for the next rice crop, and the residual phosphate fertilizer applied to the soybeans is readily available to the rice crop.

After World War II a hungry world needed rice, the export demand was brisk, and domestic production increased rapidly. By 1955 the world demand had leveled off, but American farmers continued to produce more rice. The government had to impose acreage allotments and marketing quotas to prevent overproduction. World demand increased once again in the early 1970s, and the government suspended marketing quotas in 1973, but it retained allotments on a stand-by basis. In 1983 rice exports dropped abruptly, and the price tag on the rice-support program suddenly tripled, from $250 million a year to more than $750 million. Some critics wonder whether the United States should be subsidizing farmers to grow rice that is immediately exported.

Rice farming is big business. The rice farmer must have expensive machinery to level the land, to pull the levees, to provide water for irrigation, to harvest the crop, and to dry it. He cannot shift in and out of rice production quickly in response to changing market conditions, and he needs protection against the fluctuations of the volatile world export market. In 1982 only 11,445 farmers in the United States grew rice, but they had an average of 282 acres each, and they exerted considerable influence on government price-support programs.

Today's principal rice-producing areas in the South are on the Coastal Prairies along the Gulf Coast of Texas and

Louisiana and in the Delta area of eastern Arkansas and adjacent states (Figure 18–1). The grassy prairies of southwestern Louisiana and Texas were used only for cattle ranches until railroads were built through the area in the late 1880s. The railroads brought farmers from the three "I" states (Iowa, Illinois, and Indiana) into southwestern Louisiana. They rented land from the local cattlemen and started to grow rice just as they had grown wheat in the Midwest, in large fields, with large machines and large grain elevators in the small towns along the railroads. They grew rice for two years, then put the land in improved pasture for beef cattle for four years, before growing rice on it again.

Farmers from Louisiana carried rice cultivation westward into southeastern Texas, where the flat, low-lying, heavy soils were difficult and expensive to drain. Canal companies acquired large tracts of land, developed extensive drainage systems, sold irrigation water, and rented land to rice farmers. Many Texas rice farmers were and still are specialized producers like Tommy Wollams, who own only their machinery and rent cropland for a season or two. They face increasing competition for their irrigation water from urban and industrial users.

The Texas Coastal Bend area southwest of Houston has the largest rice farms in the South. There are only about 800 of them, but in 1982 they grew an average of 450 acres of rice, with an average yield of almost 5,000 pounds an acre. The average rice farmer in Louisiana and Arkansas grew about 230 acres of rice and harvested about 4,500 pounds an acre. Rice yields do not vary much from year to year because they are influenced only slightly by weather conditions, but over the long run the trend in yields is definitely upward.

Farmers began to grow rice on the Grand Prairie of eastern Arkansas, southeast of Little Rock, in 1904. The

Grand Prairie is a broad flat terrace of older alluvium between the valleys of the White and Arkansas rivers. The brown silt loam soils of the Grand Prairie are underlain by a heavy plastic clay that holds water on the surface. The land is too poorly drained for cotton, which does not like to get its feet wet, but it is excellent for rice because the clay subsoil prevents loss of water by percolation when the fields are flooded.

In the 1950s rice farmers expanded their operations to newly cleared areas in northeastern Arkansas and to the Delta country of Mississippi. They leased land that was too wet for cotton and grew such good crops of rice that local farmers began to follow their example. The suspension of acreage restrictions on rice in 1974 triggered a rapid expansion. The rice area in Arkansas and Mississippi increased from .5 million acres in 1972 to 1.5 million acres in 1982, while the rice acreage in Louisiana and Texas increased only slightly.

Stuttgart, Arkansas, is in the heart of the Grand Prairie. The rice-drying-and-processing plants in Stuttgart have an assemblage of grain elevators that can be matched by few places anywhere in the world. The telephone directory lists 34 aerial spraying services and 19 chemical companies that serve the needs of rice farmers.

The countryside around Stuttgart is flat and empty. Low earthen levees enclose the fields and snake across them, but there are no fences. The only trees are near farmsteads, which are a mile or more apart because the farms are so large. The farmsteads are clusters of hulking, angular, corrugated metal sheds, some no more than a roof supported by sturdy poles. Each farmstead is a veritable machinery depot, with tractors, trucks, plows, scrapers, combines, and irrigation gear all over the place. Here and there in the fields a low corrugated metal roof shelters a wellhead, where a powerful diesel engine pumps water

for irrigation into long straight surface ditches or underground pipes.

The countryside is dotted with reservoirs of various sizes
and shapes. Most farmers rotate rice with two years of soybeans, but some rotate rice and reservoirs. After a year or
two of rice they flood the field for a year and stock it with
catfish and bass, but fish droppings may enrich the soil so
much that the following crop of rice is susceptible to lodging: it is easily beaten to the ground by heavy rain and
strong winds, and it is difficult to harvest.

Triple T Farms is nine miles east of Stuttgart at the edge
of Little Lagrue Bayou. The three T's stand for John Ed
Tarkington, 36, his father, Ed, 71, and his grandfather,
John, 95. When I showed up at his farm in 1982 John Ed
was welding a support bar for a four-row chisel plow he
was mounting behind a levee puller. After he had pushed
up his welding hood I joked about the money his talents
could make in the city, and he said, "That's exactly why I
do the job myself on the farm. I can't afford to pay city
prices to have it done for me. I detest paying for what I can
do myself, and I do as much of my own work as possible.

"I love to make things with my hands, and welding is my
hobby. I'm constantly modifying and improving standard
machinery, and I like to design and build my own equipment. I take a lot of pride in it. It would cost $1 million if it
was all new, but the current assessment is $675,000. I have
nine tractors—one four-wheel drive, two 180 horsepower,
two 150 horsepower, one 100 horsepower, and three
smaller ones—plus a bulldozer and an industrial backhoe.
And sometimes I think I'm in the trucking business instead
of being a farmer. I have four semitrailers, two bob trucks,
and three smaller trucks, plus the pickups."

His equipment could handle 3,000 acres, but he was
farming only 1,950, of which 1,500 were cropland. He said
he could take on 500 or 600 more acres if it was close and

the right kind of ground, but he would lose control if he got any bigger than that. His farm was above average for the area, but he said it was the right size for him. He could do things a 500-acre man could not do, but he would lose efficiency if he had 1,000 more. He owned 920 acres and rented the rest from three different owners (Figure 18–2). He rented land on a 15- to 30-year lease with complete management rights. The owners could ask him why he was doing something, but they could not tell him what to do.

John Ed had a firm three-year rotation with rice the first year, soybeans followed by winter wheat the second year, and soybeans the third year, which gave him 500 acres of rice, 1,000 acres of beans, and 500 acres of wheat on 1,500 acres of cropland. He began to plant rice around April fifteenth. During the growing season he had to maintain a flood on the rice fields—four inches of water on the deepest parts, at least an inch and a half on the shallowest. He would pump a field full and then try to hold the flood for several weeks to keep down pumping costs. At the lower ends of some of the fields he had relifts that pumped the water back to the reservoir so he could use it again.

"I love what I'm doing," he said. "Every day is different. I do construction work, I'm a mechanic, I drive machines. I do irrigation, harvesting, and laser surveying. I also do some custom surveying for my neighbors. I clear land, and I have more trucks than most fleet operators. It's a great place to live and raise a family, but the stress is tremendous. When the crops are in the ground you have to worry about everything—the weather, grass, bugs, diseases, muskrats, blackbirds, the market.

"Farming is a bigger gamble than shooting craps at Las Vegas. I've been farming for fourteen years, and I figure I'm doing all right if I can make money two years of every five. I can take a loss of $20,000 to $30,000. It's the interest

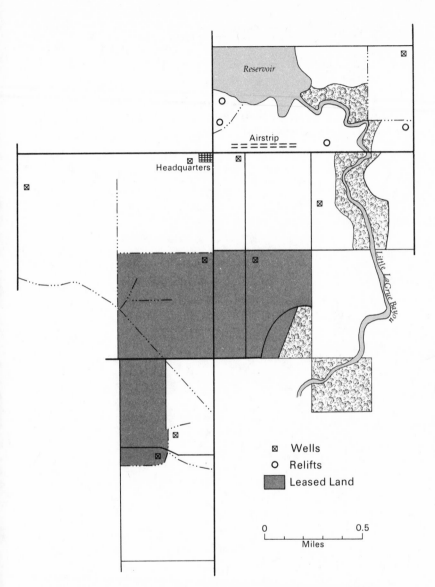

Figure 18–2. Triple T Farms. The three T's are John Tarkington, 95; his son, Ed, 71; and his grandson, John Ed, 36. The wells
raise water for flooding the fields. The relifts at the lower ends
of fields pump water back to the reservoir to permit its reuse.
The airstrip is for crop dusters.

payments on the loss that really hurt you. The last three years have been bad, but this year is the worst I've ever seen. A lot of tenants have run out of equity. I'm soaking up my equity and gambling that the next two or three years are going to be good."

19

The Sugar Bowl

Sugarcane is yet another traditional specialty crop that is produced in the cropland islands of the South (Figure 19–1). One of the two areas in which it is grown is southern Louisiana. The French-speaking people who began to settle in southern Louisiana after 1700 clearly understood the facts of live on floodplains. They knew that a river creates its bottomlands when it floods and that the floodplain belongs to the river. People can rarely resist the temptation to cultivate the fertile alluvium that the river has deposited on its floodplain, but they do so at their own risk. Sooner or later the river will rise again and overflow its banks, despite the best efforts of people to control it with dikes and levees. The flooding river will inundate the bottomlands and imperiously sweep away the works that people have so laboriously constructed, but the people will return again and again, like ants coming back to a nest that some malevolent soul has destroyed.

The highest and best-drained parts of the bottomlands are the natural levees (the banks right next to the stream), where the river dumps the greatest part of its load of sediment each time it overflows. The land slopes ever so gently away from the natural levees toward the back-swamps, which even today are still heavily wooded because they are so difficult to drain. The French divided the land into long narrow strips, more or less at right angles to the river, to give each settler some of the better levee land and some of the poorly drained backswamp. The customary farm was a long lot, at least eight times as deep as its frontage on the river.

The land is so flat and so low-lying that the settlers had to drain it before they could cultivate it, and farmers still must struggle constantly to maintain their drains. Long, straight, open ditches slice across the fields every 150 to 250 feet to carry water from the levees down to the back-swamps. Smaller ditches, called quarter drains, join the main drains at right angles every 500 to 1,000 feet. The quarter drains divide the fields into "squares" or "cuts." The farmer must clean out his ditches each time he cultivates his fields, and he must battle the rank growth of weeds that flourishes along them.

The first white settlers in southern Louisiana, who came directly from France or from the French West Indies, developed large plantations along the Mississippi River as far

Figure 19–1. Sugarcane areas in 1982. The principal sugarcane-producing districts in the South are on the levees of southern Louisiana and in the Everglades area of southern Florida.

north as Baton Rouge. Their first cash crops were indigo and tobacco. They planted sugarcane as early as 1742, but sugarcane did not become an important crop until after 1794. The planters copied the sugar-making system of the West Indies almost exactly. They planted the same kind of cane, modeled their mills after those in the West Indies, imported slaves to work the land, and even built the same kinds of houses.

A sugar planter needed a sizable acreage of land, large gangs of workers, an abundant supply of capital, and considerable managerial skill to grow and process sugarcane successfully. Many plantations had their own sugarhouses, where the juice was squeezed from the stalks and boiled down to make raw brown sugar. A tall water tank towered above the sugarhouse, and close by was a gangling derrick for handling the great masses of cane stalks that were hauled in from the fields.

Near the sugarhouse were the quarters for the workers, a line of identical, one-story, weatherbeaten wooden shacks facing a dusty unpaved road. The quarters remained compact settlements even after slavery was abolished because sugarcane production required gang labor and, unlike cotton, the crop could not be produced successfully by individual sharecroppers on small acreages. Many plantations had stores for their workers, and some of the larger ones had schools and churches.

A bit removed from the hurly-burly of daily operations, but still close enough to ensure close supervision, was the elegant mansion of the planter, set back on a grassy lawn dappled by the shade of live oak trees dripping with long gray beards of mysterious Spanish moss.

The population of southern Louisiana gained a second French-speaking element in 1755, when British forces expelled the Acadians, or Cajuns, from Nova Scotia (the land they called Acadia) and transported them on warships to the swamps west of New Orleans. The Cajuns settled in long lots along the Bayou Lafourche, Bayou Teche, and other distributaries of the Mississippi River. The Cajuns, a cheerful, friendly, easy-going people who know how to enjoy life to the fullest, could see no point in killing themselves by hard work. They raised enough crops to live on and supplemented their diet and their income by hunting, fishing, and trapping in the swamps. They are still predominantly rural, and they are patronized by the urban Creoles of New Orleans, whose ancestors came from France or from the French West Indies.

The Cajun settlers put their roads on the higher, drier land along the winding levees, and they built their houses next to the road and the river. They had large families, and their descendants divided their land down the middle, at right angles to the river, making it narrower and narrower over the generations. Many farms are little wider than the farmstead, but one to three miles deep. The houses are crowded so close together in continuous strips along the levees that southern Louisiana seems to have some of the longest streets in the world.

Sugarcane had already become the leading crop on most of the suitable land in southern Louisiana by the time Americans began trooping in after the Louisiana Purchase in 1803, and the Americans had to settle for what they could buy from the earlier French settlers. They were never able to acquire much acreage in the Cajun areas, where land was precious because families were large and farms were small.

During the nineteenth century planters gradually pushed sugar production northward toward the Delta, especially in the years when cotton prices were low. Eventually they went beyond the climatic limits of the cane plant, and they were forced to pull back. In 1982 sugarcane production was concentrated west of New Orleans, south of

Baton Rouge, and east of Lafayette on the natural levees along the Bayou Lafourche, the Bayou Teche, and the south bank of the Mississippi River (see Figure 19–1). In Louisiana they call this part of the state the Sugar Bowl. Farmers grow sugarcane on more than two-thirds of the cultivated land in the Sugar Bowl, and cane is the only crop of any consequence.

Sugarcane is a tall, thick-stemmed tropical grass with long, drooping leaves. It needs 12 to 15 months to reach full maturity, but farmers in Louisiana, even in the very southermost parts, must expect a killing frost any time between Christmas and Valentine's Day. They want to leave the crop in the field as long as possible, to obtain the greatest possible growth, but they also must harvest it before an early freeze turns the sugar sour.

The sugarcane plant grows remarkably rapidly during the warm summer months—an inch or two a week in April and May, six inches a week in June and July, and three a week in August and September. It stores sugar in its stalks, which are an inch in diameter and eight feet tall by October. A sugarcane stalk has annular nodes every six to eight inches. Each node has a bud, or "eye," that sprouts and produces a new stalk if it is covered with moist soil. Farmers use the best stalks from each year's crop to plant the following crop. They work the land into broad ridges a foot or so high and six feet apart, open a furrow down the center of each ridge, place stalks in the furrow, and cover them with soil. The first shoots emerge in two to three weeks, and normally they sprout secondary shoots to give a dense stand of cane in the rows.

The plants regenerate naturally after they have been harvested, and each planting normally produces two or three crops. The second and later crops are called stubble crops. After harvest the farmer normally plows up only his poorest fields. He cultivates them regularly during the summer to control weeds, then plants cane on the clean ground in August and September before the next harvest gets under way.

The sugarcane harvest begins in the middle of October and continues through December. The farmer uses a special cane-harvesting machine that slices off the immature top of each plant, cuts the stalk near the ground, strips off the leaves, and piles the stalks in rows behind the machine. He burns off the dry leaves and trash by setting fire to the rows with a tractor-mounted butane lighter like a giant cigarette lighter. Then he picks up the stalks with a grab loader and places them in a tractor-drawn wagon, which either hauls them to a field transfer station or directly to the sugar mill if it is close enough.

Sugarcane stalks are long, awkward, and bulky. Both the farmer and the mill must have derricks, loaders, and other heavy equipment to cope with them. An acre of cane produces about 20,000 stalks. The harvested stalk is six to eight feet long, and it weighs about three pounds. A stalk is about 80 percent juice and the juice is about 10 percent sugar, so each stalk contains about a quarter of a pound of sugar.

The stalks must be processed as quickly as possible because their sucrose starts to break down the moment they are cut. Mill operators allocate their processing capacity by assigning each farmer a daily quota for the weight of cane he will be permitted to deliver to the mill. The harvest season stretches out for several months because farmers cut only enough cane each day to make their quotas.

At the mill the stalks are sliced, shredded, and crushed between rollers that squeeze out all their juice. Lime is added to the juice to precipitate its impurities, and the clarified juice is heated to evaporate part of its water and turn it into a thick syrup. The syrup is boiled under a vacuum to crystallize the raw sugar, which is light brown because the crystals have a thin film of brown molasses that must be removed at a sugar refinery. The thick dark syrup from which no more sugar can be crystallized is called blackstrap molasses. It is a nutritious feed for cattle, and it can be distilled to make ethanol or rum. The fibrous residue of the stalks is called bagasse. Most of the bagasse is used to fuel the mill furnaces, but it can also be used to make paper and fiberboard.

Sugar enters a veritable jungle of competing products, political squabbles, and government programs when it leaves the refinery. Cane sugar is produced in 80 different tropical countries. Four-fifths of the world's sugar is consumed in the countries where it is produced, but the rest is exported, and many developing countries depend on sugar as a major source of foreign exchange. Cane sugar produced in the tropics has to compete with sugar made from sugar beets grown in the middle latitudes in developed countries, and not even a chemist can tell whether sugar was made from cane or beets.

Sugar beets were first grown for strategic reasons by countries that wished to ensure their supply of sugar in case a wartime blockade cut off their imports from tropical areas. Sugar-beets growers must be subsidized heavily, and lavish subsidies have encouraged them to produce large surpluses in some European countries. These countries also subsidize exporters to enable them to compete on the world market with the cheaper cane sugar grown in the tropics. Sugar beets surely are a leading contender for the world's most absurd and unnecessary crop: sugar from beets costs more to produce and competes with the cane sugar produced in developing countries in the tropics that desperately need the money.

Sugar products also have to compete with noncaloric sweeteners, such as saccharin and aspartame, and since 1970 they have been undersold by high-fructose corn syrup, which has partially replaced sugar in soft drinks. American per capita consumption of sugar remained remarkably stable at 100 pounds from 1950 until 1973, but it dropped to only 70 pounds in 1984. Consumption of corn sweetners increased from only 10 pounds in 1950 to 25 pounds in 1973 and to 50 pounds in 1984.

Unlike other countries, which have only a single source of sugar, in 1982 the United States produced 2.8 million tons of cane sugar, 3.2 million tons of beet sugar, and 3.2 million tons of high-fructose corn syrup. It also imported 3.0 million tons of sugar. Any national sugar policy must balance the demands of cane growers (in Louisiana, Florida, Hawaii, and Puerto Rico), beet growers (mainly in irrigated sections of the West), corn growers, and friendly, underdeveloped countries that need our dollars. It is further complicated by the wild fluctuations of the world price of sugar. Long periods of low prices are punctuated by sudden explosive increases every five to ten years. The world price for a pound of sugar, for example, was 2¢ in 1968, 30¢ in 1974, 8¢ in 1978, 29¢ in 1980, and 4¢ in 1984.

The object of our national sugar policy has been to protect domestic growers and to ensure that a substantial share of the nation's sugar requirements is produced within our borders. The secretary of agriculture allocates quotas to domestic areas and to friendly foreign countries that are discontinued when the world price is high and reinstated when it drops. Consumer groups complain that our sugar programs have added $2 billion a year, or about $10 a person, to the price Americans pay for their sugar.

It really is not surprising that sugar programs in the United States are complex and controversial at a time when other countries are subsidizing increased production, the world has a growing surplus, and the world price is well below the domestic price. In 1982 the average world price of raw sugar was 8.4 cents a pound and the average domestic price was 19.9 cents a pound. The average cost of producing a pound of raw sugar was 11.1 cents in Louisiana and 13.4 cents in Florida, where labor costs were higher.

Most sugar-beet farmers grow beets on relatively small acreages in a three-to-five-year rotation with other crops. They might be able to shift to less remunerative crops, such as alfalfa, corn, soybeans, wheat, sunflowers, or potatoes. Sugarcane, however, is a monoculture crop that is grown on large acreages, and cane growers have no ready alternative. The loam soils of the natural levees in southern Louisiana could grow vegetables, but our ability to consume vegetables is finite, the demand is already adequately supplied, and developing satisfactory marketing facilities for vegetables would be extremely difficult and expensive.

In 1982 Arthur J. Bergeron, the county agent in Assumption Parish, Louisiana (where counties are called parishes), told me, "Only a limited amount of land in southern Louisiana can be cultivated. We've got about 40,000 acres of cane in this parish, and that's about it as far as our good land is concerned. The rest is called rice land, but we don't grow any rice. Every now and then somebody puts a crop of soybeans on it, but it's so wet he can't harvest it. Some of it is used for crawfish ponds.

"Most farmers here don't want to grow anything but cane, and they've been doing it ever since granulated sugar was developed in the 1700s. Sugarcane has been a dependable crop in this area. It's not a hard crop to produce. You've got to cultivate it and control the weeds and fertilize it, but a lot of the time not much is happening. You need 300 to 500 acres for a reasonable cane farm. What's the going price for land? Shoot, they ain't none for sale! A fellow that needs more land will rent it for a fifth of the crop.

"You need at least four big tractors and two harvesters, just in case one breaks down. They cost $80,000 to $100,000 and you only use them for two and a half months a year. You do repair work on the machines while you're waiting to start harvesting because you can't afford to have one break down on you in the middle of harvest. Then you need field wagons and a transloader to transfer the stalks from the wagons to the truck that will haul them to the mill where they are crushed. We had 54 mills in Louisiana back in 1953, but now we're down to only 21" (Figure 19–2).

Bergeron suggested that I visit U. B. Simoneaux, whose farm is across the Bayou Lafourche from Paincourtville, about four miles north of Napoleonville. The road out to his farm wound along the levee. On one side was the bayou, 20 yards wide and 20 feet below the level of the road. Household refuse, empty beer cans, and more mysterious matter drifted in its murky, stagnant water. The other side of the road was an almost continuous line of buildings, ranging from trailers and narrow shotgun shacks to modern ranch homes, with an occasional monumental Catholic church. The land sloped away almost imperceptibly toward the wooded swamps a mile or so off in the damp, hazy distance. The chocolate-colored soil had been corrugated into ridges a foot or so high that bristled with low, bushy clumps of young sugarcane plants. Water stood in the furrows between the ridges.

U. B. Simoneaux told me that his family has been in this area for more than 150 years. His grandfather had a plantation with his own sugar mill. U. B. owns 180 acres and works 356. Cash rent in this area runs about $25 an acre, but he leases land for one-sixth of the crop. Most of it is in long narrow strips (Figure 19–3). He leases three of the strips from the three Edwards brothers. "I don't know how they did it," he said, "but they managed to make a living on their three little silvers of land.

"We probably ought to get bigger. You should have at least 500 acres if you want to stay in the cane business, but you can't pick up anything right now. I think I can make it on this acreage if the prices stay, and I'm not attracted by the management necessary on the larger places. You spend all your time riding around in a pickup truck. I have two full-time black laborers, four during the grinding season. One is 64, and the other is over 70. Both of them live nearby in their own houses.

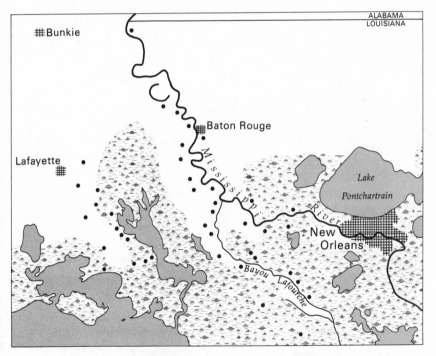

Figure 19–2. Sugar mills in Louisiana in 1970. The mills are on the natural levees along the Bayou Teche south of Lafayette and along the Mississippi River and the Bayou Lafourche south of Baton Rouge. Much of the low-lying backswamp area between the bayous has never been drained.

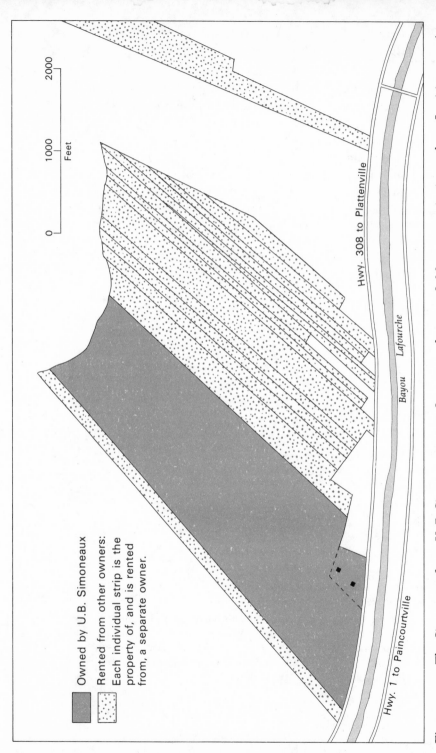

Figure 19-3. The Simoneaux farm. U. B. Simoneaux owns 180 acres and rents 176. Most properties in southern Louisiana consist of long, narrow strips that have been laid out years or less at right angles to the general course of the principal stream.

Owned by U.B. Simoneaux

Rented from other owners:
Each individual strip is the property of, and is rented from, a separate owner.

Hwy. 1 to Paincourtville

Bayou Lafourche

Hwy. 308 to Plattenville

Feet
0   1000   2000

"Last year we grew 256 acres of cane for the mill, and we left 100 acres fallow for a year. It's traditional to leave it fallow, but we probably ought to plant something on it. We've tried wheat and beans, but they didn't work out too well. In August we cut our very best cane and plant the stalks on the land that's been fallow. It takes one acre of cane to plant four acres, and even then we probably haven't been planting enough. The cane keeps stubbling, but we usually plow it up after three years. We keep it in stubble if it looks like it will make 25 tons or better. Last year we got 34 tons per acre."

U. B. starts harvesting in the middle of October and keeps going until after Christmas. He delivers 90 tons a day to the mill during the grinding season. The mill will send out trucks, but it pays him $2 a ton to haul it himself. During the summer he has to cultivate the fields to control weeds, and he has to clean out the cross drains every time he works a field. "Last year I didn't get around to it once," he said, "and we had a new lake on the place."

20

River of Grass

The other major sugarcane-producing district on the mainland of the United States is the Everglades area south and east of Lake Okeechobee in southern Florida (Figure 19–1). The Everglades are the southern half of a gentle trough of low-lying, poorly drained swamp and marshland that reaches 250 miles down the center of the state from Lake Tohopekaliga just south of Orlando to the Gulf of Mexico.

Lake Okeechobee is in a large, shallow basin near the center of the trough. In its natural state the lake had no clearly defined shoreline. The edge of the lake was where the water was too deep for plants to grow. The lake simply overflowed along much of its southern rim, and a sheet of water 40 to 70 miles wide seeped southward through the Everglades to the Gulf of Mexico. The land slopes no more than two or three inches to the mile, and the water oozed along at a rate of only about 20 feet a day. It was only a few inches deep during the winter dry season, but heavy rains in summer and early fall could raise it as much as two feet or more.

Southern Florida is one of the wettest areas in the United States. The rainfall varies greatly from year to year and from season to season, but it averages about 60 inches a year. Most of the rain falls between June and October. Summer days are hot and muggy, and the moist air is so unstable that thunderheads boil up nearly every afternoon. Every other summer day has a short but heavy thundershower, and two or three inches of rain can pelt down within an hour or so. In the fall tropical hurricanes bring torrential rainstorms that pound the earth for several days. A single hurricane can drench the area with more rain in two days than falls on any part of Iowa during an entire summer.

Winters are mild and relatively dry. The alternation of wet summers and dry winters has encouraged the growth of sawgrass, a tough, coarse plant that is not really a grass at all, but a sedge that grows six to ten feet high. A vast sea of gently waving sawgrass, which stands in water for half the year or more, once covered the entire area. The Seminole Indians, who took refuge in the area and learned to navigate it in their canoes, called the Everglades "the river of grass."

The remains of the sawgrass plants gradually accumulated when they died and fell into the water because the normal microorganisms of decay require oxygen, and they cannot work on vegetative matter that is under water. The residues of partially decomposed marsh plants in the Everglades form one of the world's largest deposits of organic—peat and muck—soils. Peat soils have at least 65 percent organic matter, and muck soils have 25 to 65 percent. The organic soils are eight to ten feet deep near Lake Okeechobee, but shallower away from the lake.

Organic soils can be extremely productive if they are drained. The soils of the Everglades are especially good because they have been formed in a basin that is floored with limestone, so they have a high lime content. They lack copper, manganese, zinc, and other trace elements that are essential for plant growth, but the trace elements are needed only in minute quantities that are easy to supply.

In 1907 the state of Florida created the Everglades Drainage District to drain the organic soils of the area.

This agency dug five major drainage canals eastward to the Atlantic Ocean and one westward to the Gulf of Mexico. The main canals are eight to ten miles apart. Private landowners have crisscrossed the area between them with a herringbone of lesser canals, drains, and ditches. Most of the farmland is divided into blocks of forty acres, half a mile long and an eighth of a mile wide, that are bordered by lateral canals six feet wide or field ditches three feet wide. Water percolates laterally through the soil both to and from the ditches, and farmers control the water level in their ditches to irrigate as well as to drain. They pump water out of their ditches and into the canals to drain the land during the rainy season, from June through October, and they can pump water from the canals into their ditches to provide water for irrigation during the winter dry season.

The construction of drainage canals had lowered the water level in the Everglades enough to enable farmers to start growing winter vegetables on a large scale when a disastrous hurricane struck the area in 1928. Lake Okeechobee is less than 20 feet deep, and steady, galeforce winds from the north blew the northern end of the lake completely dry. The winds drove the water to the southern end of the lake, where they lashed it into waves more than ten feet high. These waves battered and finally breached the low muck levees, and the water poured through in a devastating flood that drowned more than 1,800 people. The Corps of Engineers was immediately called in to build a sturdy new dike 38 feet high around the lake. The canals drain the agricultural area, and the dike keeps the lake from overflowing and flooding it ever again.

The drainage of the organic soils of the Everglades has created valuable agricultural land, but it has also generated a whole new set of problems. The soil begins to subside as soon as it is exposed to air. The microorganisms of decay attack the vegetative matter, and the soil is so spongy that it is easily compressed by heavy farm machines. After a long dry spell it may even catch fire and smolder for days until it is soaked by a heavy rain. The soil is disappearing at a rate of about an inch a year, and soil subsidence may force the abandonment of much of the agricultural land by the year 2000.

Farmers can minimize subsidence by keeping the water table as high as possible when the soil is cultivated and by flooding it when it is not. They like to keep the water table low, however, because a water table that is too low harms their crops less than one that is too high, and a low water table is also a precaution against sudden heavy rains. Pasture is the most protective form of land use because a sod cover and a high water table slow the rate of subsidence, but pasture also gives the lowest returns. Most farmers in the Everglades will tell you that their land is too valuable to be "wasted" under pasture.

The most important crops in the Everglades are sugarcane and vegetables. Some vegetable farmers grow sugarcane as a sideline crop, but more than two-thirds of the cane is produced by four large corporations that grind their cane in their own mills. In 1981 the U.S. Sugar Corporation, the largest, owned 168,000 acres of land and grew sugarcane on 125,000 acres. It operated two sugar mills and maintained a huge fleet of harvesters, field wagons, tractors, trucks, and other equipment. It employed 2,400 regular workers, many of whom lived in ten company villages, and it hired an additional 5,500 seasonal workers. It ran beef cattle on land that was not well suited to cane production and fed them blackstrap molasses from its sugar mills.

The acreage of sugarcane in the Everglades was fairly small until 1960, when the United States placed an embargo on imports of Cuban sugar after Castro took over

(Figure 20–1). Farmers had grown sugarcane in the Everglades on a small scale in the 1920s, but as late as 1959 the Everglades had only 47,000 acres of cane. Five years later the sugarcane area had increased to 214,000 acres, and in 1982 it was up to 344,000 acres, almost half of the national total.

The sugarcane harvest in the Everglades runs from November through April. The rich, black muck soils support such lush growth that the mature stalks curve in all directions instead of standing erect, and no one has been able to develop a machine that can harvest the tangled stalks satisfactorily. The best machine available is self-propelled on caterpillar treads. Long arms protrude awkwardly to the front and one side. The front arm has revolving blades that slice off the tops of the stalks. The machine then grabs the stalks, cuts them into short lengths, and puts them on a conveyor belt in the side arm that transfers them to a trailing field wagon pulled by a four-wheel-drive tractor.

Cane growers harvest as much as they can by machine, but 70 percent of the crop still has to be cut by hand. The cane cutters are housed in dormitories and hauled to and from the fields each day in battered, old school buses. A field is fired the day before it is cut to burn off the dry dead leaves and other trash and to drive out any poisonous snakes before the cutters start work. During the harvest season the distant skyline will suddenly erupt with a great billowing cloud of dense black smoke, but the fields burn quickly, and all that remains, by the time one can find a way to a field through the maze of drainage ditches, is the rich, caramel fragrance of burnt sugar and a tangle of blackened stalks with a few leaves that are still smoldering.

The workers cut the stalks with vicious-looking machetes, hack off the tops and any remaining leaves, and toss the stalks behind them. After the field has been cut, a

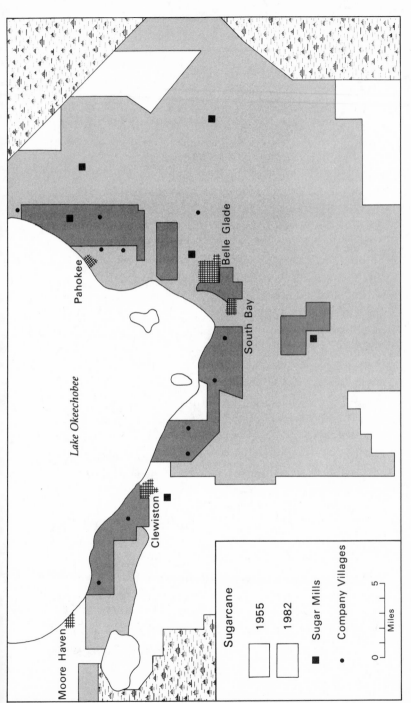

Figure 20–1. The Everglades sugarcane area. The acreage of sugarcane was fairly small until 1960, when the United States placed an embargo on imports of sugar from Cuba. Much of the cane is grown by large corporations that have their own company villages to house their workers.

self-propelled loader comes rumbling in on caterpillar treads, scoops up the stalks, and loads them into angular field wagons with wire mesh sides that can hold four tons of cane. Four-wheel-drive tractors haul trains of four field wagons to a transfer ramp near the field. At the ramp each wagon in turn is tilted to dump its contents onto an angled conveyor belt that loads the stalks onto the massive cane trucks that haul twenty-ton loads of cane to the mill be to ground.

The sugar companies need a dependable supply of temporary workers for the dirty and unpleasant job of harvesting the fire-blackened cane. Most of the cane cutters are imported from the sugar islands of the British West Indies. The sugar companies complain that even though they advertise and recruit vigorously, pay good wages, and offer attractive fringe benefits, they still are forced to import cane cutters because "American workers have not demonstrated their willingness, ability, or availability to cut cane in Florida."

The farmers who grow winter vegetables in the Everglades also need a dependable supply of temporary workers to harvest their crops, but they rely on local people rather than on migrants (Figure 20–2). Winter vegetables are second only to sugar as a source of farm income in the area, which has the mildest winters on the mainland of the United States. January has an average low temperature of 55° F, and daytime highs run around 75° F. Every other year has a night of frost, when the temperature drops as low as 32° F, but only one year in five has a freeze, when the temperature drops low enough to kill plant tissue, and cold spells rarely last more than a night or two.

Cold air masses from the north are warmed when they cross Lake Okeechobee, and frost protection is greatest southeast of the lake near Belle Glade, which boasts that it is the winter vegetable capital of the world. In 1982 Ken

Shuler, the county agent, told me that farmers in the Everglades grow more than 40 different vegetable crops.
"They grow sweet corn," he said, "leafy vegetables, such
as celery, lettuce, escarole, endive, parsley, cabbage, and
cauliflower, and root-type vegetables, such as radishes,
carrots, and onions. The farmers on the sandy land over
near the coast grow fruit-type vegetables, such as peppers,
tomatoes, eggplant, beans, and squash.

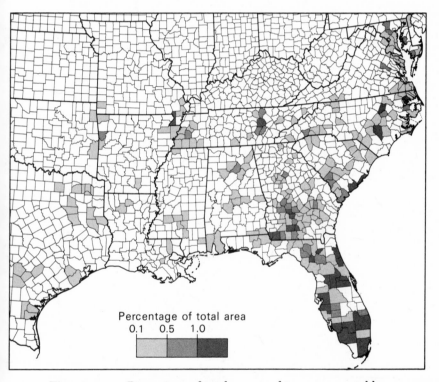

Percentage of total area
0.1   0.5   1.0

Figure 20–2. Percentage of total area used to grow vegetables
in 1982. Vegetables were grown fairly widely throughout the
South, but the leading areas were in central and southern
Florida. The Everglades area south of Lake Okeechobee is one
of the nation's leading winter vegetable producing districts.

"A vegetable farmer needs at least 50 acres to use modern machinery efficiently," he told me, "but most of the growers in the Everglades have 1,000 to 10,000 acres in vegetables. The larger growers may employ 350 people year-round and more than 2,000 at the peak of the harvest. Most of them own their land, but they rent fields for a year or two from a cane company that wants to get rid of cane diseases, and they grow cane on their own land to control vegetable diseases."

Most vegetable growers specialize in a few crops, but they also produce others. They start planting in September, and they are still harvesting in April or May. Unlike most farmers, who grow only one crop a year, the vegetable men grow several crops a year on the same land. They can grow seven crops of radishes, two of lettuce, maybe two of sweet corn. They stagger their planting dates to spread the harvest, even out their labor needs, reduce frost risk, and catch the highly volatile market.

The vegetable land is fallow from May through August because the summers are too hot and too wet for growing vegetables. Vegetables can stand heat, and they can stand water, but they cannot stand both at the same time. The combination of 85° F temperatures and soggy feet will steam them to death. The growers flood their fields after the last harvest to reduce oxidation of the soil and to control pests. The Everglades have a good climate for growing winter vegetables, but its problems with weeds, plant diseases, insects, and other pests are aggravated by the lack of a sustained cold season and by the abundant moisture supply during the warm months.

Shuler suggested that I talk to Ray Roth, who had built up a successful vegetable farm from scratch. "I was born in Cleveland in 1924," he told me. "My father had a truck farm that was annexed into the city before I started school. He grew just about every kind of vegetable you can think of, so I learned how to grow everything. When I got out of service I wanted to build a greenhouse, but I couldn't borrow the money. In 1948 I came down here to visit a service buddy, and the next year I came back and sharecropped with his boss. I had ten acres of endive and escarole."

In 1950 he leased 35 acres, and in 1951 he leased 40 more. He gradually built up to leasing 640 acres in 1957, and in 1962 he and two other fellows tied up 11,000 acres at what is now the home farm. A couple of Cubans had taken an option on the land to grow sugar on it. Ray and his partners picked up the option when the Cubans left to join the Bay of Pigs operation. They bought the land as a corporation and sold it to themselves for $100 an acre average. "It was raw land mostly," Ray said, "that had never been farmed, all sawgrass and willows. You could just reach out and grab a handful of mosquitoes. It had been drained in 1925, and they tried to grow bananas on it, but got froze out."

Ray bought 1,600 of the 11,000 acres. He paid $300 an acre for it because it was ready to farm, but he still had to put in roads and ditches. He applied 25 pounds of copper to the virgin soil, but has never had to add any more. By 1964 he had paid for the whole 1,600 acres. He has bought more land when it has come on the market and when he could afford it. Land sells for $4,000 to $5,000 an acre near the lake, where the climate is milder and the soil is deeper, but farther from the lake the price may be as low as $1,800.

"The price of land also depends on how long it has been cultivated," Ray said, "because this organic soil has a life, and some day it just ain't gonna be there no more. The soil was six to eight feet to lime rock when I bought it, but now it's only 30 to 36 inches because of oxidation and compaction. The particle size used to be an eighth of an inch, but now it's just like flour. We think we have at least another 20 years, but that's what they were telling us 20 years ago.

"We needed new land because we were having problems with diseases and soil compaction on the land that had been in vegetables for a while. It's not a good idea to grow the same vegetable crop on the same land for too long because of diseases, and you need to take the land completely out of vegetables for a while. I'd like to raise vegetables for three years and then to go sugarcane for three years. The main reason for my last land acquisition was to get a better rotation of vegetables and cane. You can't make big money in cane, and it's no fun because it's too easy to grow. We plant and cultivate cane, but the co-op handles the harvesting and marketing."

In summer Ray Roth has to flood the vegetable fields to keep down weeds and to reduce soil loss by oxidation. About ten years ago he started growing rice in the flooded fields as a summer cover crop, and plowed it under in the fall to add fiber to the soil, which it does not get from vegetables. Four years ago he realized that there was a large market for rice among the Spanish-speaking people in Miami, and since then he has been combining and selling the grain.

Although he grows sugarcane and rice, Ray is basically a vegetable man. He says they take more work, more capital, and more personnel, but they are also more profitable. He has made as much as $3,000 an acre from single crops of radishes, carrots, and lettuce. He grew sweet corn when he started, but now he is entirely in radishes and leaf vegetables. He has no vine crops, only salad crops. He grows 800 acres of leaf crops, 400 acres of radishes, 200 acres of carrots, and 140 acres of parsley. Yes, 140 acres of parsley! He gets two crops a year from each field, except when they are in radishes, which produce six or seven crops a year.

His farm headquarters east of Belle Glade has two machine sheds with shops, two rice bins with driers, and a setup for washing radishes. He owns two $87,000 tractors,

an $80,000 rice combine, a $50,000 carrot bagger, a
$30,000 radish bagger, two custom-built ten-row radish
harvesters and one custom-built six-row carrot harvester
that cost $50,000 apiece, a power shovel for cleaning
ditches, and an awesome array of equipment for cultivation and irrigation.

Ray starts preparing his land for crops in August, and he
staggers planting to give him something to harvest every
week. He cultivates the root crops to control weeds, but he
sprays herbicides on his leaf vegetables instead of cultivating them because soil is almost impossible to get out once
it gets inside the leaves, and he said that dust storms in
winter can spread enough dirt as it is without any help
from him.

He harvests radishes and carrots by machine. The radish
harvester digs up ten rows at a time, trims off the tops, and
loads them onto a conveyor belt that carries them to a
trailing truck. He gets 300 to 400 boxes from each crop of
radishes and carrots. How big is a box? "It varies with the
crop," he said. "A box of parsley holds 60 bunches, and a
box of endive 18 to 24 head. The vegetable growers have a
committee that's been trying to come up with a standard-
sized box for vegetables for six or eight years now, but it
hasn't gotten anywhere."

He harvests the leaf crops and parsley by hand. He gets
about 500 boxes from each crop. He does everything possible on a piecework basis—hoeing, thinning, harvesting,
and loading. He has 28 full-time employees and hires more
than 175 people in the harvest season. They all live in the
area, and they go on welfare when there are no jobs for
them. At the peak season, from December through February, he has to meet a payroll of $50,000 a week.

"My operation is incorporated," Ray said. "We gross
around $5 million a year. Of course I spend time on the
farm, but I'm not really a farmer any more. I have an out-

standing manager, and he runs the farm. I spend a lot of time trying to keep finance lined up, and I spend time at the packinghouse, talking to the salesmen, finding out what they are able to sell. One of the most important things in agriculture today is marketing, and I attribute our success to the skill of our marketing organization."

He said he had three choices in selling vegetables: he could take his chances selling at the packinghouse in Belle Glade, where buyers come from the chain stores, as he did when he first started; he could take his chances selling on consignment, shipping to commission merchants in major cities who sell if they can, but that is pretty uncertain; or he could sell FOB (free on board), as he decided to do in the 1950s when he joined the South Bay Growers Co-op. The co-op was formed to give growers a more rounded package for marketing. It sells 40 different vegetables, and can fill one truck with different kinds. "Have you ever tried to sell a whole truckload of radishes or parsley?" Ray asked. "All of the sales are made over the phone by salesmen who live here. They maintain telephone contact with buyers all over the North. Last year they sold $50 million worth of farm produce, all of it by telephone. Our sales force is so good that basically we only cut what we have already sold.

"The cost of harvesting is the only thing that matters in the produce business. You've already paid the cost of seed and fertilizer before you can harvest the crop, but harvesting is the big expense. You don't harvest a crop unless you can sell it for enough to break even. If the price is too low you just plow it under and plant the block again. A lot of growers figure they're doing all right if they harvest two-thirds of what they plant, but over the years we have left less than ten percent of our crops in the field.

"None of my crops can be frozen or processed. They must be sold immediately. All the produce is hauled to the packinghouse I own in South Bay, where it is cooled, graded, and packaged. Today everybody wants a uniform quality package, and we strive to produce the very best. We deliver the packaged produce to the co-op platform, and it is trucked to wherever it has been sold by the sales force."

"Vegetables are very easy to overproduce. This is a major supply area, and we control the total amounts produced, so a disaster here is reflected in our prices. The prices go up in a bad year because supplies are down. If California controls the supply we can have a bad year and also get a poor price. Of course a good year here can also hurt you because everybody has an abundant supply, and the price goes down. We have absolutely no government interference whatsoever, no quotas, no allotments, no price supports, nothing like that. The vegetable growers in Florida like it that way!"

21

The Citrus Ridge

Most Americans probably think of oranges when they think of Florida, and rightly so, because the state produces three-quarters of the nation's oranges, and oranges are its most valuable agricultural product. Oranges were introduced to Florida by Spaniards, who planted citrus trees soon after they founded St. Augustine in 1565. Seminole Indians carried fruit from the Spanish settlements to all parts of the peninsula, and they scattered the seeds so widely that later visitors assumed, quite incorrectly, that oranges were native to the state.

The first commercial orange groves were planted in 1819 in the northeastern part of the state near navigable waterways such as the St. Johns River. The early groves were small. The fruit was packed in barrels padded with Spanish moss and shipped in slow sailing schooners to ports in the Northeast. Oranges arrived during the winter holiday season when other fruits were in short supply. They were such an expensive luxury that many children saw them only once a year, when they found oranges tucked in the toes of their Christmas stockings.

In the winter of 1894–95 severe frosts devastated the citrus groves of northeastern Florida. The owners replanted their groves immediately, but in 1899 another extremely severe freeze killed most of the tender young trees. The citrus business in the area never recovered. Growers moved south and began to develop new groves on the warm, sandy ridge down the middle of the state that had been made accessible by the completion of a rail-road from Jacksonville to Tampa in 1884 (Figure 21–1).

The central ridge is the principal citrus-producing area in Florida. The greatest concentration of groves is west of Orlando and east of Lakeland. U.S. 27, the Citrus Trail, passes close to two-fifths of the nation's citrus groves be-tween Ocala in the north and Frostproof in the south. An-other fifth are near U.S. 1 north and south of Fort Pierce in the Indian River country along the Atlantic coast. Most

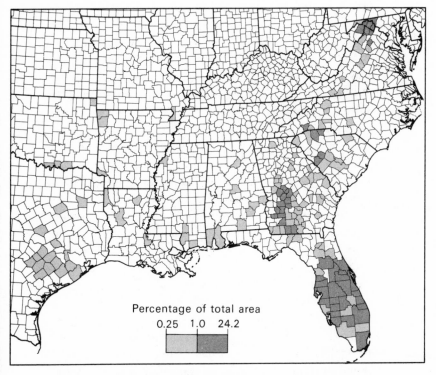

Percentage of total area

0.25   1.0   24.2

Figure 21–1.   Percentage of total area used for orchards in 1982. Oranges and grapefruit are the principal orchard products of Florida. The orchard areas in southwestern Georgia produce pecans and peaches.

Florida oranges are processed into frozen juice concentrate, but the Indian River area specializes in premium-quality fresh fruit for direct consumption. It also has one-third of the nation's grapefruit trees. The warm, moist climate of Florida is best suited to producing juice fruits, such as oranges, grapefruit, and tangerines, but lemons do better in the hot dry areas of California and Arizona.

The citrus ridge is the backbone of Florida. It is a low, lake-speckled area that is underlain by limestone. The highest parts are called "mountains," even though they are little more than 300 feet above sea level. The limestone is extremely soluble. It is honeycombed with caves and subterranean passageways, many of which are occupied by large underground rivers that emerge in giant springs. The ridge is pockmarked with thousands of sinkholes, depressions where the underlying limestone rock has been dissolved. The groundwater table is so high that many of these depressions contain lakes.

During the recent geological past, when sea level was several hundred feet higher than it is today, the limestone ridge was covered by thick deposits of sandy material. The porous, gray, sandy topsoil is extremely droughty, and it contains almost no major plant nutrients. "This citrus land is so high and dry that it's not much good for anything else," said Robert E. Norris, the county agent in Lake County, Florida, when I visited his office in 1959. Lake County is one of the state's leading citrus counties.

"We produce fruit from fertilizer," Norris said, "not from the natural fertility of the soil. The soil doesn't do much more than hold up the trees, and we supply all the plant food they need. We select grove sites for frost protection, not for the fertility of the soil, because it doesn't have any to speak of."

Summers in central Florida are hot and muggy, but the

winters are pleasant. The average daytime temperature in January runs around 70° F, but every winter has a couple of cold spells of two or three nights each. The worst danger of frost is in the last two weeks of December and the first two weeks of January. Fruit is damaged if the temperature drops to 28° F for four hours, and temperatures of 20° F or lower for eight hours can kill the trees (Figure 21–2).

A freeze is caused by a cold, dry air mass from the north. The first night, when the cold front passes, is fairly windy, but the second night is cold and still, with clear skies. On such cold, clear nights the soil radiates heat to the atmosphere and cools rapidly. The cold soil chills the air next to it, creating a temperature inversion. The layer of air near the ground is colder and heavier than the air higher up, and it drains downslope to the lower areas, which are frost pockets. Frost damage is least likely on the ridges and the higher slopes and most likely in the low-lying areas where cold air accumulates.

Water bodies hold heat much better than the soil. Lakes give frost protection by moderating the temperature of the cold air that crosses them and by warming the cold air that accumulates above them. A slope on the southeastern side of a large, deep lake is the best location for a citrus grove.

Citrus grove owners start their groves with young trees that have been top-grafted at a nursery. Spring is the time to transplant. A young grove should be fertilized every six weeks and older groves three times a year. All groves must also be sprayed for insects and diseases at least three times a year. A grove needs a summer cover crop to protect the light, sandy soil from the heat of the sun and from erosion by torrential thunderstorms. In the fall the grove owner plows under the cover crop to add organic matter to the soil and to allow the bare soil to absorb heat from the sun.

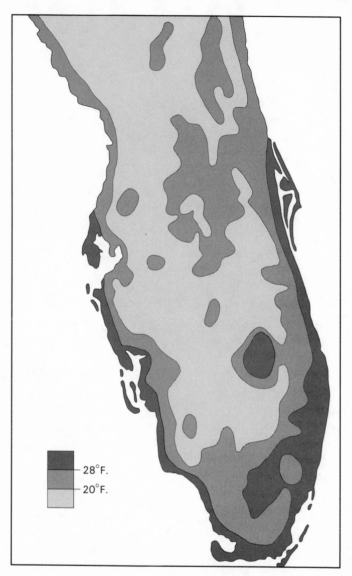

Figure 21–2. Temperatures in southern Florida at 0600 on January 12, 1982, showing the moderating effects of water bodies. Areas along the coast were warmer than areas inland. The numerous lakes along the Citrus Ridge kept the central part of the state a few degrees warmer, and Lake Okeechobee had an obvious effect on neighboring areas.

The young trees should start to bear fruit by the fourth year, and they should begin to pay for themselves by the eighth.

Picking starts in October. A busy season in December and January is followed by a lull in March and then another peak season in May. Most of the fruit is picked by crews on a piecework basis. In the morning, as soon as the dew has dried, they lean lightweight ladders against the trees, clip the fruit from the branches, and put in it big bags that are slung over their shoulders. When the bag is full they empty it into a field box. A tractor hauls the full field boxes to the edge of the grove, where the fruit is transferred to a truck or trailer that carries it to the packinghouse.

Packinghouses specialize in producing attractive fresh fruit of high quality for direct consumption. They are huge buildings with long conveyor belts that carry the fruit through batteries of machines where it is washed, dried, waxed, polished, graded, and packed for shipment. Most of them have their own sidings where they can load boxes of fruit directly onto railroad cars for shipment to cities in the North. Some of them are independent, but some are owned by groups of small grove owners who have formed cooperative marketing associations.

Before World War II most of the fruit was shipped fresh, but by 1960 more than three-quarters of the oranges and half of the grapefruit were processed into canned juices, segments, or frozen concentrates. Production of frozen orange juice concentrate jumped from 226,000 gallons in 1946 to more than 44,000,000 gallons in 1952. In the processing plant the juice is extracted from the oranges, evaporated in a vacuum at a temperature of 60° to 80° F, flash frozen, and stored at a temperature of minus 10° F. The peel, pulp, and seed from the processing plant are ground and dried to make a good feed for cattle. The availability of citrus by-products encouraged the growth of the beef-cattle business in Florida.

"The organization of citrus production also has changed completely," Norris said. "You aren't going to find one single family-operated citrus farm in the whole darned county because the small grove owner simply cannot afford the machinery and equipment he needs. An operator has to work at least 100 to 125 acres of grove, but the average ownership is only 40 acres. Our citrus growers now hire their grove work done by custom operators that run specialized grove caretaking services. The cooperative marketing associations had to form their own production departments, which really are grove caretaking services, because so many of their members couldn't afford the equipment to produce fruit of the quality they needed.

"Most of our citrus grove owners are professional people who figure grove land is a good investment. They hire caretaking services to do all the grove work, and the processing plant handles the picking and marketing. More than half of our grove owners don't even live in the county. They put the management of their groves completely in the hands of one of the cooperative marketing associations. The co-op charges them for each grove operation, but the charges aren't billed until the fruit is sold, so their only contact with the grove is their fruit payment checks. It's almost like owning stock in General Motors or some other big corporation and getting an annual dividend check."

Five citrus growers formed the Lake Region Packing Association in 1909 as a cooperative marketing organization to pack, ship, and sell fresh citrus fruit in northern cities. By 1959 it had grown to 297 members. Half of the members lived within 35 miles of its packing plant, a quarter lived in other parts of Florida, and the rest were scattered all the way from Las Cruces, New Mexico, to

Middletown, Connecticut. They owned 4,400 acres of bearing groves and 2,200 acres of young trees in seven different counties.

The headquarters of the association was a one-story, white, concrete office building in Tavares, Florida, the county seat of Lake County, some 35 miles northwest of Orlando. The hulking packinghouse across the street covered two city blocks. Its two-story shell of gray corrugated metal sheltered three separate packing units with a total daily capacity of 10,000 ninety-pound boxes, enough to fill twenty railroad boxcars. The boxes of fruit could be loaded directly onto railroad cars on the siding along the north side of the plant or onto trucks from large loading platforms on the south side. The fruit that was not sold fresh was sent for processing at the Plymouth Citrus Products Cooperative, a co-op formed by 15 packing associations.

In 1936 the association was having trouble getting its members to produce the quality and quantity of fruit it needed, so it established a production department to provide the grove caretaking services necessary to produce top-quality fruit. The production department was a mile west of Tavares. A chain-link fence six feet high, topped by three strands of barbed wire, enclosed the ten-acre compound. Don Kemp, the production manager, had his office in a one-story, light green, cinder-block building at the entrance. The compound had two enormous equipment sheds, several hundred feet long, that were open on all sides. Slender steel poles painted rust-resistant orange supported corrugated metal roofs that fended off rain and sun from a bewildering clutter of tractors, sprayers, hedgers, irrigation guns, and other equipment. All the machines were painted red. A bevy of smaller, enclosed buildings housed workshops and facilities for storing fertilizer, insecticides, and other chemicals. All of the major equipment had two-way radios to keep in touch with its operators who might be anywhere in seven counties.

Don told me that the production department took care of 90 percent of the acreage that belonged to members of the association "from first planting until the fruit is ready to be picked. The association also has separate picking and packing departments, with a peak employment of 800 workers. I have 150 men working year-around in the production department and more than 300 during the growing season or when there is danger of a freeze. The packinghouse employs more than 200, and the picking department more than 300, at the peak harvest season. The same men may work for all three departments at different times. The association has several housing areas for its workers, including a concrete-block dormitory that can house 150 men.

"A member of the association can do all of his own grove work, or he can rely completely on us. Or he can use another caretaking service and use the association only to pick, pack, and market his fruit. His only contact with the grove may be the check he gets when the fruit has been sold and the expenses have been deducted. I have been caretaking some groves for nigh onto twenty years, and I have never once, to my knowledge, laid eyes on the owner. Of course any time I see a car with a New York license plate parked beside the grove it could be the owner."

I left Tavares in 1959 with mixed feelings about the innovative management system that had been developed in the Florida citrus country. The new system enabled a family to retain the ownership of a small grove, even when they could not afford the equipment necessary to care for it properly. Could the idea of hired caretakers, I wondered, be transferred to other areas where the soaring costs of machinery were forcing farmers to enlarge their operations? Was it the wave of the future in American agriculture? On the other hand, I also wondered about the efficiency of an operation that had to manage 6,600 acres of grove land scattered over a distance of 60 miles.

I went back to Tavares in 1982 eager to discover what new ideas had been developed about grove management and wondering whether a more efficient grove layout had been worked out. Don Kemp had retired, and John Hey had replaced him as production manager for the Lake Region Packing Association. "Things haven't really changed all that much around here since you talked to Don back in 1959," John told me to my no small disappointment.

"We still cultivate, irrigate, spray, fertilize, hedge, and top in the same old way," he said. "Forty to 50 percent of the groves are irrigated. We use more herbicide sprays, and we do less hand hoeing to control grass and weeds. I only have 25 to 30 full-time workers in the production department because we're using more machinery and less hand labor. We're still oriented toward fresh-fruit production, and we shoot for 60 percent. The rest goes to the processing plant to be frozen or canned.

"The association now has only 124 members, but their groves are bigger, and they own around 5,000 acres of bearing trees. Half of them still live within 35 miles, 20 percent live in other parts of Florida, and the rest are scattered all over the eastern United States. We mail them their checks for fresh fruit within three months of harvest, but the cannery check takes about a year."

In 1983 Florida had a big freeze. There was nothing unusual about that because Florida seems to have a big freeze nearly every winter, and the initial reports of damage usually are pretty exaggerated. There were freezes in 1977, 1978, 1980, 1981, and 1982, for example, and after each one those who enjoy making dire predictions had a field day making dire predictions about the gloomy future of the citrus industry in Florida, but before long it was business as usual once again in the groves.

Grove owners cannot really assess the severity of a freeze until two or three months afterward when new growth should start to appear, but by that time the news media have long since lost interest. The grove owner has to prune away the dead wood and bulldoze and replant the dead trees, but he or she accepts these as part of the normal cost of doing business. In 1982 I could find precious few signs of damage from the freezes that had occurred in two of the three previous winters.

The Christmas freeze of 1983 was quite another story. It absolutely devastated 200,000 acres of citrus groves, an area one-sixth the size of the entire state of Delaware (Figure 21–3). Two years later the groves north of Orlando still looked like they had been cauterized by a forest fire. The long, straight lines of trees were rows of gnarled, gray ghosts. Their trunks and branches were gaunt and bare, without sign of leaf or life. The dead black wood showed through cracks where frost had split the bark. Lush green weeds flourished beneath the trees where spraying had been discontinued. An occasional plume of dirty smoke marked the funeral pyre of a pile of dead trees that a bulldozer had uprooted and heaped for burning. Flocks of snow-white egrets picked over the worms and insects in the soil overturned by the bulldozers.

I stopped at the Lake County agricultural extension office in Tavares to ask John Jackson, the citrus specialist, about the freeze. "The temperature dropped from a daytime high of 80° F to a low of 17° F the next night," he said, "and with it we had a thirty-mile-an-hour wind. Not many plants can stand that kind of change. We actually had even colder temperatures in 1981, but there was less damage because the change was not as sudden and we had much less wind. The warm spell caused the sap to rise in the trees, the icy wind froze it, and it expanded until it broke the bark.

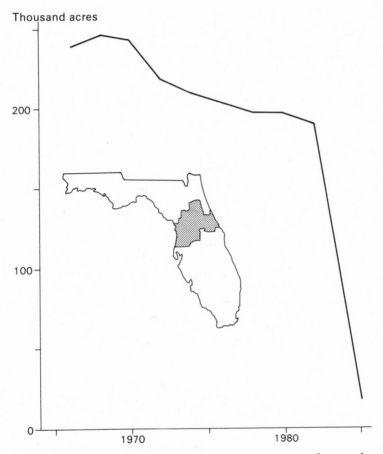

Figure 21–3.   Citrus acreage in twelve counties in north central Florida between 1966 and 1985. Grove acreage had been declining slowly under pressure of urban encroachment, but it was decimated by the Christmas freeze of 1983, which devastated more than 200,000 acres of citrus grove land.

"Our citrus acreage has been declining slowly anyhow,"
he said, "because central Florida is one of the fastest-grow-
ing areas in the United States. Some growers have been
bulldozing their groves for urban development, but
freezes have hit the citrus business pretty hard. In 1962
'the freeze of the century' dropped us from 27 million
boxes to 13 million. We were up to 45 million boxes by
1980, but in 1985 we produced less than half a million. In
1982 Lake County had 117,730 acres of citrus, but in 1985
we only had 12,183 acres, and that was just what was alive,
not necessarily bearing. It's going to take us at least ten
years to get back to 15 million boxes."

I suggested that the combination of urban development
and freezes might force the citrus business to relocate in
less vulnerable areas, but he disagreed vigorously. "We've
surveyed grove owners about their plans," he said, "and
half of them expect to replant within three years. A quarter are undecided, and the rest are going to do something
else with their land. Thousands of acres of new groves
have been planted in the last 10 or 15 years in areas east
and south of the Citrus Ridge, especially by large citrus-
processing corporations, but many of the new groves are
on poorly drained land that is going to be difficult to manage. We can still produce top-quality fruit in this area, and
I think we'll come back strong with fresh fruit."

Can Florida compete with Brazil, which began large-
scale production of oranges around 1980? Brazil now produces as much citrus fruit as the United States and exports
most of it as concentrate. The Brazilians first began to
compete with us in our traditional export markets, but
then they began to ship directly to this country in competition with domestic producers. In 1980 they had only
twelve percent of the U.S. market for juice concentrates,
but their share had jumped to nearly half by 1984. The
quality of some of it is not as good, and it has to be blended

with American juice, but the amounts threaten growers in Florida, who produce most of the U.S. concentrate.

American consumers used to resign themselves to having to pay more for their morning orange juice every time they read about a freeze in Florida, but Brazil has begun to fill the gap. Recently the price of orange juice has risen for only a week or two after a Florida freeze before concentrate from Brazil starts flooding in to lower it again. William Wintersdorf, executive vice-president of the Lake Region Packing Association, has warned his members that "we must remember that we are, and have been for several years, part of a world market that is being supplied, to a greater extent than ever before, by fruit-producing sources outside the continental United States. Florida is not the domestic or world supply source that it once was for fresh or processed citrus."

22

Too Much
Good Land

American agriculture has become increasingly international since World War II. Before 1940 most American farmers produced only for domestic consumption, and they could ignore the rest of the world. The value of agricultural imports regularly exceeded the value of exports, but the imports did not compete with domestic products because they were mainly coffee, bananas, spices, and other tropical crops. By 1980, however, American farmers had become keenly attuned to the world market because the United States exported two-fifths of all the crops it produced. In 1981, the peak year, we exported agricultural products worth $43 billion.

The United States has more good farmland than it needs to feed and clothe its people, so American farmers must export because they can produce more than the American people can consume. The surplus may be either a vexatious problem that must be controlled or a blessing that should help to feed a hungry world, but American farmers suffer when other countries, for whatever reason, do not or cannot buy the surplus that they have produced.

The United States is the supplier of last resort to which other countries turn only when they have no alternatives. The world market plays crack-the-whip with the supplier of last resort. This supplier prospers when the world suffers a poor crop year because demand is strong and prices are good, but it accumulates huge surpluses when the world enjoys a good year, and its prices are depressed accordingly.

Farmers demand government action when prices are poor. A basic function of government is to protect society, or significant segments thereof, against the hidden hand of the market, and American farmers have not hesitated to invoke governmental protection when a weak market has depressed the prices they have received, even though the market is weak because they have produced more than the rest of the world can use or more than it can afford. They have forced the government to make up the difference between the market price and some higher "fair" price they have negotiated through the political process.

Most governmental farm programs in the United States have been attempts to cope with the chronic surplus resulting from the ability of our farmers to produce more than the market can consume. American farm policy has shuttled between two conflicting philosophies: 1) support prices by restricting production to what the market can consume; or 2) encourage production and exports, and let the world market determine the domestic price. Should one have greater confidence in the government or in the market?

Both approaches have serious flaws. Tight controls on production might ensure good prices for farmers, at least for a while, but they would require elaborate supervision. They would also raise prices for consumers, which could make the American market so attractive to foreign producers that the United States would have to erect high tariff barriers to protect American farmers against competition from cheaper imports. Conversely, dependence on the world market could cause severe hardship for farmers when it was glutted, and they might properly demand government subsidies to smooth out its peaks and troughs.

In the 1920s Congress responded to agricultural surpluses by passing bills that were intended to support farm prices by restricting production, but Presidents Coolidge and Hoover vetoed them. The Agricultural Adjustment Act (AAA) of 1933, which became the basis for all subsequent agricultural legislation, was signed into law by President Franklin D. Roosevelt as part of his New Deal. The AAA attempted to raise prices by reducing production, and a generation of Americans remembers Secretary of Agriculture Henry A. Wallace as "the man who killed all the baby pigs."

Agricultural overproduction was not a problem during World War II, when American farmers were urged to produce as much as possible to support the war effort, or immediately after the war, when they were implored to help feed a hungry world. Surpluses began to build up, however, and in 1954 the Agricultural Trade Development and Assistance Act (Public Law 480) authorized the export of surplus farm commodities to poor countries at subsidized prices for emergency relief, for blocked foreign currencies, or for strategic materials.

Public Law 480 clearly had a high humanitarian objective, but it also served the practical political purpose of helping to reduce the growing surplus of farm products. Other humanitarian attempts to reduce surpluses include the food-stamp program and the school-lunch program. One of the more effective nonhumanitarian programs was the Soil Bank, in existence from 1956 to 1975, which at its peak removed 58 million acres from agricultural production and facilitated the retirement of large areas of marginal farmland.

Surpluses built up during the 1960s, and prices were not good, but the early 1970s were banner years for American farmers. The easing of international tensions encouraged exports of farm commodities to the USSR, which increased from $0.045 billion in 1971 to $1 billion in 1973 and to $4 billion in 1979. New markets opened in China late in the decade. Rising incomes in countries such as Japan shifted consumption from rice to wheat, soybeans, and meat, which required Japan to import feed grains. Devaluation of the dollar enabled Third World countries to borrow cheap money to import food. American agricultural exports exploded from $7 billion in 1970 to $41 billion in 1980.

World prices for crops were high, and American grain farmers responded enthusiastically. High crop prices hurt livestock producers, who needed feed for their animals, and high food prices fueled domestic inflation, but farmers eagerly grew more corn, wheat, and soybeans. Rising costs of production cut their profit margins per bushel, so they squeezed more bushels out of each acre, and they acquired and farmed more acres.

The experts urged farmers to enlarge their operations. "Borrow money to buy land and machinery," said the experts, and they ridiculed any farmer who was not in debt as a conservative old stick-in-the-mud. Bankers and other lenders also urged farmers to borrow money, and they offered low interest rates. Farmers were competing with each other for land, and they quickly bid up the price, using the escalating price of their land as collateral for the money they borrowed to expand. High crop prices were built into the price of farmland. Farmers were not making much money from farming, but the rising price of farmland made them feel prosperous. Euphoric lenders assumed that the good times would never end and that the price of farmland would continue to escalate. They made loans to farmers even when they knew the farmers could not make enough money to repay them. Farmers used their land as their collateral, and the lenders were sure they could sell it for a tidy capital gain if ever they had to foreclose on the loan.

During this period the price of farmland was rising faster than the rate of inflation, so it attracted nonfarm speculators, who could borrow money to buy farmland at interest rates below the rate of inflation. Favorable tax laws encouraged wealthy people to use farmland as a tax shelter: they could improve the land and deduct their expenses from their nonfarm income; then they could sell it at an inflated price and pay a low capital gains tax on their profits.

As late as 1980, when worrisome farm surpluses had already started to build up, newspaper financial pages were still touting farmland as a good investment, and alarmists were even fretting about an impending shortage of farmland. The National Agricultural Lands Study predicted that U.S. food exports would triple between 1980 and 2000, and they concluded that the conversion of farmland to nonagricultural uses could create serious food shortages by the turn of the century. In 1982 Crosson and Brubaker estimated that the United States would need at least 60 million acres of new cropland in the next thirty years. The 1982 Yearbook of Agriculture, *Will There Be Enough Food?*, sounded the same alarm.

It was already becoming apparent, however, that the 1970s had been a golden age for American agriculture and that the golden years were ending. American farmers had tooled up to supply a market that was rapidly disappearing. The crash came in the early 1980s, when the bottom dropped out of the export market. The United States had mortgaged its future by cutting taxes while spending huge sums of money to buy new toys for the generals and admirals. Staggering budget deficits forced the dollar sky-high and priced American farm products out of the international market. A global recession reduced the ability of poorer countries to buy the food they needed, and they were forced to cut their imports drastically. American corn exports declined in 1981, wheat exports dropped in 1982, and soybean exports followed in 1983.

The loss of export markets depressed domestic prices in the United States. The prices of farm products came tumbling down, and the prices of farmland came tumbling after. In southwestern Minnesota, for example, the average price of farmland soared from $844 an acre in 1975 to $2,083 an acre in 1981, but then plummeted to only $967 an acre in 1985.

Falling land prices undermined the value of the collateral against which farmers had borrowed money to expand, and falling crop prices reduced their ability to keep up their payments. The young, aggressive, risk-taking go-getters, who were praised as the best and most progressive farmers in the 1970s, were in the deepest trouble in the 1980s. In the 1970s some people had been begging them to produce more food to feed a hungry world. In the 1980s these selfsame people were denouncing the farm expansion of the 1970s as "a wild spending spree" by farmers.

The declining value of farmland made lenders nervous. Some farmland was worth little more than the loan that was outstanding against it, and the lender who had to foreclose on the loan and sell the land could easily lose money. Lenders in farming areas raised their interest rates to cover their losses on bad loans and to protect themselves against risky new loans. They even were reluctant to lend money to farmers for seed, fertilizer, fuel, and other normal production expenses.

American farmers must somehow recapture the export markets they lost in the early 1980s, but it will not be easy because farmers in other countries have greatly expanded their output. The world food crisis has shifted from shortage to surplus. The United States has been too successful in helping Third World countries to become self-sufficient in food production, at least from the viewpoint of American farmers. Countries once on the verge of starvation now are feeding themselves and even producing a surplus. India suffered a famine in 1965–1967, but exported nearly a million tons of rice in 1981. The new surpluses could be wiped out overnight by bad weather, insects, diseases, or other disasters, but in the long run it looks like American farmers may remain the suppliers of last resort to whom other nations turn only in time of distress.

Some of our former customers have become keen competitors. They subsidize exports heavily because it is better, they say, to sell food even at cut-rate prices than to let it rot in warehouses. First they undersell us in our traditional export markets, and then they export directly to the United States and undersell us here. In April, 1985, the *Des Moines Register* reported that soybean oil imported from Brazil was sold in New Orleans at prices lower than soybean oil produced in Iowa and shipped down the Mississippi River to New Orleans by barge.

Government farm programs often backfire when they try to reduce surpluses, restrict imports, or increase exports. In 1983, for example, the United States developed a Payment in Kind (PIK) program that paid farmers to idle part of their land by giving them crops from surplus stocks in lieu of the crops they would have grown themselves. The farmers could pocket the money they would have spent to grow the crop. The PIK program did little to reduce surpluses, but it devastated the sales of implement dealers, seed and fertilizer merchants, and other small-town businesses that depended on farmers as their customers.

An attempt to formulate farm policy for an area as large and diverse as the United States will almost inevitably have unintended side effects. The goal of the dairy whole-herd buy-out program in 1986, for example, was to reduce milk production but maintain the price of milk by buying out the least efficient milk producers and slaughtering their entire herds for beef. Eliminating the least efficient producers hardly made a dent in the milk surplus, but the program threatened cattlemen. Dairy cows are not good for much except hamburger, but dumping tons of hamburger on the market would have depressed the price of beef and wreaked havoc with cattle ranchers in the West, who were just beginning to recover from two years of severe drought and what had seemed like an eternity of low prices.

The bane of one farmer is the boon of another. A drop in the price of corn, for instance, will hurt the corn farmer, but it will help the farmer who buys corn to fatten cattle, hogs, or poultry. Back in the old days, when farmers sold a variety of commodities, they could shift their emphasis to the one that was most profitable at any given time, but greater specialization has made farmers increasingly vulnerable to changes in the prices of the limited number of commodities they produce.

Farm policy is based on legislation that is extraordinarily complex and detailed. It is enacted by Congress, under pressure from legions of lobbyists, who know far more about its implications for their specific commodities than any poor, overworked Congressman could ever hope to learn. Trying to keep track of all the convolutions and particulars of farm policy is a full-time job, and few people endure the drudgery unless they have strong and compelling reasons to do so.

For example, the Food Security Act of 1985 (popularly known as the 1985 farm bill) runs to more than 300 tightly packed pages of legalese in fine print. It includes a provision (Sec. 1771) directing the secretary of agriculture to convey a tract of 0.303 acres of land in Irwin County, Georgia, to the county Board of Education. Perhaps the good members of the Irwin County Board of Education were dancing in the streets after the bill had been signed into law, but who else even knows where Irwin County is or why the tract was transferred? And who cares?

After Congress has passed a farm bill it is implemented by voluminous regulations promulgated by bureaucrats in the U.S. Department of Agriculture. Then lower level bureaucrats must try to explain the maze of regulations to individual farmers, who immediately start looking for the loopholes. Farmers like to joke about "farming the government," but actually they are quite serious because their livelihood depends on it. There simply is no way that a bunch of politicians and bureaucrats in Washington is ever going to be able to stay ahead of American farmers for very long.

The goals of the 1985 farm bill were to increase exports by allowing prices to drop to world market levels and to maintain farm income with subsidy payments based on loan rates, target prices, deficiency payments, and acreage set-asides. The loan rate is the price the government guarantees a farmer for his crop. The farmer expects to forfeit the "loan" and let the government add the crop to its surplus stock unless the market price rises above the loan rate.

For each crop the secretary of agriculture sets a loan rate at a price that is supposed to be competitive on the world market, and Congress sets a target price that is supposed to give farmers a fair return on the crop. The government makes up the difference between the loan rate and the target price by giving the farmer a direct cash subsidy, which is called a deficiency payment. At the end of 1986, for example, the loan rate for corn was $1.80 a bushel, the target price was $3.03 a bushel, and the farmer received a deficiency payment of $1.23 for each bushel of corn he produced.

In order to qualify for deficiency payments farmers

must enroll part of their land in acreage set-aside programs that are intended to reduce surplus production by
cutting back on the amount of land that is cultivated. The
government assigns each farmer an acreage base for each
crop. The farmer agrees to plant the crop on only part of
this base acreage, and the government pays him the average value of the crop he would have grown on the land he
has set aside.

The initial results of the 1985 farm program were not
encouraging. The prices of crops dropped in the United
States and our exports were cheaper, but other countries
lowered their own prices to match or beat ours, which
kept exports low and surpluses high. Some observers have
counseled patience. They argue that the federal budget
has more impact on agricultural exports than anything
that might be included in a farm bill. They believe that
exports will pick up when the value of the dollar declines
and the health of the global economy improves.

Other observers argue that we should concentrate on
the economic good health of American farmers and to
blazes with the world market. They say we should maintain domestic farm prices by restricting production to the
amount we need in the United States. In the short run
such an effort might succeed, but in the long run it would
be doomed. High prices in the United States would inevitably attract imports from other countries, and the high
tariff barriers necessary to exclude them would invite ret-
ribution and damaging trade wars. Some people believe
that the Smooth-Hawley Tariff caused the Great Depres-
sion, and they do not want an instant replay. Eventually
the price of food in the United States would rise so high
that consumers would revolt, and farmers would be in hotter water than ever.

The choice between a policy of dependence on the
world market and a policy of government regulation and

protection is partly a matter of political ideology, but the only sure guarantee of high prices in the United States is the guarantee that they will eventually put American producers out of business. In the modern world no single nation or small group of nations can long dictate an inflated price for any commodity.

The history of cotton shows how American farmers can shoot themselves in the foot by trying to maintain high prices for a crop. Between 1924 and 1928 the United States produced three-quarters of the world's cotton and supplied nearly two-thirds of the world's exports (Table 22–1), but the price of cotton was so low that cotton farmers demanded a higher price. The guaranteed price of cotton in the United States became an "umbrella" over the world price. It encouraged farmers in other countries to start growing and exporting cotton. By 1978–1983 the United States had been reduced to producing less than a fifth of the world's cotton and supplying only a third of its exports, and the nation's cotton acreage had shifted from the Old South to districts in the irrigated West where it was cheaper to produce.

The American position in soybeans in 1979–1983 was

	U.S. Percentage of World Production	U.S. Percentage of World Exports	Percentage of U.S. crop that was exported
1924–1928 average			
Cotton	74	64	59
1979–1983 average			
Cotton	18	33	55
Soybeans	62	82	42
Corn	43	78	30
Wheat	15	43	61

Table 22–1   U.S. Share of Specific Crop Markets

comparable to the nation's cotton position between 1924 and 1928 (Table 22–1). The price in the United States is the price that farmers in other countries have to beat in order to undersell American farmers, and foreign soybean growers positively drool at the prospect of a high guaranteed price in the United States. Have we learned the lesson that cotton should have taught us, or will we allow high prices to wipe out our soybeans too?

Low prices are merely a symptom of the basic problem: American farmers have become entirely too good. They can grow twice as much as their fathers grew on the same land or the same amount on only half as much land. They are producing prodigious surpluses of virtually all agricultural commodities, but the world no longer needs the food and fiber they can produce in such rich abundance.

Excess farm capacity in the United States is a chronic condition, not merely the temporary aberration that some observers seem to hope it is. Much of the nation's farmland is surplus to present and foreseeable needs, and it must be retired from agricultural production. The question is not whether, but how. Will it be retired willy-nilly, the way cotton land was abandoned in the South, or can it be retired in some orderly, planned, and rational manner? How can public policy facilitate the process? What can we do to alleviate the pain and personal suffering that will be inevitable if the government tells farmers that they must stop farming some or all of their land?

The United States needs a bold, innovative, and massive land-retirement program that will remove from agricultural production some of our best farmland and all of our good, marginal, and environmentally fragile land. The retirement of our less productive and most vulnerable land is simple common sense when we are producing so much more than we can use or sell, but the retirement of large acreages of marginal land unfortunately will not make much of a dent in our agricultural surpluses because such land, by definition, does not produce very much. Some of our best farmland must also be retired and placed in reserve against the day when it will be needed once again.

Such a prescription may sound fairly drastic until you remember the experience of cotton in the South, which could easily be repeated with other crops in other parts of the United States. The market will do the job for us if we do not have sense enough to do it for ourselves first. The United States can produce crops at competitive prices if our crop production is concentrated on our best land in our most productive areas. Our other farmland should be retired. We need to take careful inventory of our land resources and to assess them unsentimentally to identify the areas best suited to the production of the food and fiber we need, can use, or can sell.

Some people fear that the retirement of farmland will harm rural communities by undermining the economy of the small towns that serve them, but farmland retirement programs have been far less devastating to rural communities than programs that have merely idled farmland temporarily. Many observers also have failed to realize that the basic function of small towns has changed. Once they were central places that served the agricultural areas around them, but they began to lose out as retail centers when the Model T Ford replaced the horse and buggy, when rural people could travel farther to get the goods and services they needed. Retailing activities have been moving up the urban hierarchy, from smaller to larger places, but employment in manufacturing has been trickling down, from larger to smaller places. Main Street is dead, but small towns thrive because they have become cogs, however small, in the national system of manufacturing centers.

A program of land retirement should be combined with

a reduction in the price of the land that remains in production to enable American agriculture to regain its lost advantage. The United States cannot hope to compete with the rest of the world in labor costs, and at the moment its principal advantage seems to lie in technology and scale of production. Genetic engineering and other emerging biological technologies bid fair to increase agricultural productivity as much as it was increased first by machinery and later by chemicals. The new technology will require large investments in basic research and development, and it will require expensive production inputs that are beyond the means of farmers in most parts of the world.

Cheap land once was one of the principal advantages enjoyed by the United States, and we should try to regain it. Reducing the inflated price of farmland should be a major goal of our farm policy because it would enable American farmers to compete more effectively in world markets. Reducing the price of farmland will be extremely painful because the value of their land is a major component of the personal net worth of most farmers. They spend their working lives building up equity in their land, just as city people build up equity in a pension or retirement plan, and they expect it to support them in their old age. The precipitous drop in the price of farmland in the early 1980s was a terrible shock to older farmers, who have seen their entire life savings melt away before their very eyes. Imagine how city people would feel if the value of their pensions were suddenly cut in half. Perhaps we could save money if we terminated all price-support programs and offered pensions to farmers for a specified number of years.

Retiring large acreages of cropland and reducing the price of the land that remains in production have political and economic costs that may be unbearably high. The costs should be one-time costs, and they should put American agriculture on a sound competitive basis that would require no further government intervention, but they will require political decisions so difficult that they may be impossible to make. What politician worth his salt is going to take kindly to telling voters that not one single acre of land in their state is worth retention for agricultural production? Our political leaders may duck the issue and leave the future of American agriculture to be decided by the harsh workings of natural economic forces.

A do-nothing policy may be encouraged by the growing ignorance about farming and the loss of support for it in our society. Urban legislators and consumer advocates have become increasingly critical of the high costs of farm programs, and some observers have begun to lump agriculture with steel, automobiles, textiles, and shoes as examples of sick, bumbling, and declining industries that have lost their ability to compete in the modern world. The romance of saving the family farm seems to be wearing thin.

23

The Family
Farm

The American family farm has become a family farm business. Once upon a time the farmer and his family grew crops, fattened and butchered their own cattle and hogs, milked a few cows, kept a flock of chickens for eggs and Sunday dinner, tended a garden and orchard, and preserved their own vegetables and fruits for the winter. All that is past. The modern farmer has had to specialize in doing what he can do best, and he has gotten rid of everything else. He no longer even tries to feed himself and his family because it is cheaper to buy food at the supermarket.

Farming in the United States since World War II has been in the throes of the same kind of transformation that revolutionized manufacturing during the eighteenth century and retail trade during the twentieth. Small-scale operations are just as outmoded on the land as they have become in the city. Some people jump to the erroneous conclusion that the new farms will be "factories in the fields." The home workshop became a factory, to be sure, but the "ma and pa" grocery store on the corner was transformed into a supermarket, and the new family farm is merely a much larger and more efficient version of the older edition.

Like the grocer, the modern farmer has had to increase the volume of his business because his profit per unit is so small. The farmer has increased the number of units he produces by increasing his yields per acre and by renting land from his neighbors to give him more acres. The average size of farms in the eastern United States increased from 100 acres in 1940 to 220 acres in 1982, and the yields of most crops also more than doubled.

The farmer has been able to increase his yields because plant breeders have given him better varieties of crops, and he has poured on the fertilizer to take full advantage of their genetic potential. He relies on a whole arsenal of chemicals to protect his crops against voracious insects and to get rid of weeds. He can handle a larger acreage because bigger and better machines have reduced his labor requirements. He can sit in an air-conditioned cab on a powerful new machine and accomplish more in half an hour than his grandfather could do in a long, sweat-stained, backbreaking day.

The modern farmer is producing too much, and he knows it, but he is not foolish enough to cut back unilaterally when his neighbors are also producing too much. They all depend on exports to an international market that fluctuates unpredictably, and they feel frustrated and helpless when they are battered by its wild gyrations.

The modern family farmer has had to become a good money manager on top of all the other skills he needs to remain in business. Most farmers have to borrow money at some time during the year simply to cover their normal operating expenses, such as the costs of seed, fertilizer, chemicals, and fuel. Many livestock producers must also borrow money to buy the lean animals they fatten and some of the feed they use to fatten them. In 1910 American farmers borrowed only 13 percent of the money they needed for normal operating expenses, but in 1983 they borrowed 48 percent. They had to learn to watch interest rates as carefully as they watched commodity prices.

The successful modern family farmer has had to become a specialist, and increasing specialization is changing the geography of American agriculture. Economics and politics determine the total amount of corn or milk or any other agricultural commodity produced, but the natural environment, history, and accessibility to markets determine where within the nation it will be produced. Specific commodities will become increasingly concentrated in the areas that are best suited to their production, while the marginal and less well-endowed areas will decline or drop out of production completely.

An inordinate amount of sentimental twaddle about the imminent demise of the family farm has been perpetrated by well-meaning but naive people who have been badly misled by official statistics. A successful modern family farm is a business that provides an adequate level of living in return for the labor of a parent and child, with a hired hand at those stages of the demographic cycle when the child is too young to be of much help or when the parent is too old. An operation must sell a minimum value of farm products each year in order to support a family. This minimum value was approximately $5,000 from 1950 to 1960, $10,000 from 1960 to 1970, and $40,000 from 1970 to 1982. An operation that sells less is a part-time farm that requires supplemental off-farm income.

The U. S. Census of Agriculture reports that the United States had 5.4 million farms in 1950 but only 2.2 million in 1982. This loss of 3.2 million farms would be truly alarming if it had actually occurred, but it is completely fictitious, because the official census definition of a farm is ridiculously permissive. Only 1.2 million of the 5.4 million operations officially classified as farms in the 1950 census sold as much as $5,000 worth of farm products, and only 0.6 million of the 2.2 million "farms" in 1982 sold as much as $40,-000 worth. In short, between 1950 and 1982 the United States lost 0.6 million family farms, not 3.2 million, as naive users of census statistics might think.

Roughly three-quarters of the operations officially classified as "farms" are undersized, part-time, hobby, and other kinds of "nonfarm" farms that do not produce enough to support a family. They require too much labor per unit of production, and they cannot provide income that is competitive with industrial wages. They require intensive specialization in order to be successful, but they do not command the capital necessary for specialization. Many of their fields are so small that modern farm machines cannot operate in them efficiently.

Some city people who dislike changes in the countryside have disguised their true intent by using "the family farm" as a shibboleth. They know little and care less about farming; they would rather deify than define what they mean by a family farm because they realize that it is a more potent political ploy to prattle about preserving the family farm than it is to try to get people exercised about preserving pleasant rural landscapes. Picturesque poverty for peasants may be fine for city folk who want to enjoy a nice Sunday afternoon drive, but these people would flatly refuse to live and labor on the kinds of hardscrabble farms they romanticize.

I am all in favor of keeping rural landscapes attractive, but landscape conservation should be advocated honestly and directly. In the heavily urbanized areas of the Northeast, for example, and in other areas where much of the farmland is marginal and will continue to be abandoned, it might make sense to treat the countryside as a museum and to pay farmers as custodians to keep it looking the way city people want it to look.

Some people have assumed, quite incorrectly, that small farms must be family farms but large farms cannot be. They have jumped to the totally erroneous conclusion that small farms must be good and large farms must be bad. They have invented the vague and amorphous concept of something called "agribusiness" as a convenient whipping boy, and they have used this term as a shibboleth to castigate farm operations larger than they like. Presumably a farm that sells more than $40,000 worth of products can be called an agribusiness, but a farm that sells less could not support a family, and many farmers tell me that the minimum in sales necessary for a successful family farm is closer to $100,000 than to $40,000.

Some people who are concerned about changes in the countryside believe that large corporate farms are gobbling up small family farms and taking over the land. The threat may be real in parts of the West, but in the East incorporation is merely a legal device for keeping the farm in the family. A partnership may be as good as a corporation to ensure the intergenerational transfer of farm assets. Most corporations that own farmland in the East are small family-held affairs created solely to avoid having to sell off a chunk of the farm for death and estate taxes when one member of the family dies.

The principal threat of corporate farming for American family farms comes from food companies in the United States, which are developing major production and processing facilities in Third World countries. In this country the big investor who wants a big return on his money cannot compete with the farm family that is willing to settle for merely making a living from the land and is prepared to exist on its equity when times are bad.

Some people are sincerely concerned about the changes that are taking place on our farms. Some are troubled by our increasing dependence on scarce and expensive energy resources, and some fear that climatic change might reduce our ability to produce food. Some fret, quite unnecessarily, about the loss of prime farmland to urban encroachment; at the present rate of encroachment the United States will run out of farmland somewhere around the year 2500. Some worry about the loss of topsoil to erosion, the contamination of air and water by agricultural chemicals, and residual pesticides on fruits and vegetables.

Some people who are concerned about changes in the countryside want to reject recent technological innovations and go back to an older farming approach that is variously known as alternative, ecological, environmentally sound, low-input, natural, organic, regenerative, or sustainable agriculture. They have publicized their efforts effectively, and they have pressed governmental agencies and experimental stations to develop their ideas, but most farmers are skeptical. Some people undoubtedly have been successful with farming systems that substitute human labor for chemicals and machinery, and their triumphs have been widely heraladed, but far less has been said about those who have tried and failed, or even about the supplementary income sources of some of those who have succeeded. Most farmers probably are not willing to reject the new technology that saves them labor.

Farmers should have just as much right as any other Americans to an acceptable style of life and an acceptable standard of living. They should not be condemned to lives of poverty and toil just to satisfy the romantic notions of affluent urban ideologues. The myth that anyone should be able to farm is an ignorant insult to contemporary farmers. It assumes that successful farming requires little more than brute strength, and it completely ignores the high degree of technical competence and the large amount of capital that are necessary for a successful contemporary farm.

The vast majority of agricultural commodities produced in the eastern United States will continue to be produced on farms that are owned and operated by families. Most of the farms I have described in this book are family operations. Many of them employ some additional labor, but the farmer is ready to roll up his sleeves and get his hands as dirty as necessary. He can do any job that has to be done on the farm, and he must be able to show a hired hand or anyone else how to do it.

I have talked about farm operators as though they were males because 95 percent of the farm operators in the United States in 1982 actually were men, but the farm wife is the linchpin of the family farm. Traditionally the farmhouse, the yard, and the garden have been her domain, while the barn, the barnyard, and the fields have belonged to the man. Wives have been responsible for homemaking and bookkeeping, but men have made most of the decisions about crops and livestock, and they have done most of the work in the fields.

The one implement with which most farmers feel the most uncomfortable is pen or pencil. They operate, maintain, and repair some remarkably complex machinery as a matter of course, but as like as not it was the farm wife who was expected to keep the books, and she could only wring her hands in despair when her husband ignored the concerns and worries of a mere woman about their financial situation.

The responsibilities of the farm wife have become even more demanding as good hired hands have become harder to find, machine power has replaced muscle power, and brains have replaced brute strength as the essential requisite in farm labor. On many farms the farm wife has taken the place of the hired hand. She must be able to step in and drive a tractor, operate machinery, haul grain, or tackle any other job on the farm when other help is not available. Many farm wives have also taken jobs in town to provide a steady cash flow that will help balance out income fluctuations on the farms, and many a farm depends on the paycheck of the farm wife for its very survival.

Angry storm clouds were looming on the horizon in 1982, when I visited most of these farms. One might ask why the farmers did not have sense enough the see these storm clouds, to sell and get out while the getting was good. The answer, of course, is that you cannot be a farmer unless you are an incorrigible optimist. Anyone who routinely suffers the worst that God and humanity have to offer cannot afford to be deterred or discouraged by a bad year or two.

Farmers have had to learn to cope with adversity. There are always storm clouds of one kind or another on the farm horizon. There had been bad years before, and there will be bad years again. The year 1982 was just another bad year, or so it seemed at the time, and not many of us realized how tempestuous those particular clouds were destined to be. Farm families, as they always have when times are bad, simply tightened their belts, trimmed their sails, battened their hatches, and got ready to ride out the storm. They cut their expenses to the bone, lowered their living standards, and prepared to live on the equity they had worked so hard to build up.

They hoped, as they always have, that sooner or later things were going to get better, and they have. The family farm has become a family business, but farming still remains a way of life as well, and those who farm the land have a deep and abiding attachment to it that they are not prepared to break.

Acknowledgments

A mere listing of the individuals and institutions that have helped me bring this book to fruition does not adequately express my appreciation for all the time and help they have given me, but a proper expression of my gratitude to them might require another book. I am genuinely grateful to:

- the John Simon Guggenheim Memorial Foundation for a fellowship and the University of Minnesota for a supplementary sabbatical leave that enabled me to do much of the fieldwork on which the book is based;

- the Commonwealth Fund Book Program, which facilitated its appearance in print;

- each and every one of the people who is mentioned in the text for the hospitality with which they received me and for the graciousness with which they answered my questions;

- Helene Friedman, Phil Gersmehl, John Hudson, Peirce Lewis, Cotton Mather, Sally Myers, Neil Salisbury, and Dick Skaggs for their support and encouragement;

- Ennis Chestang, Larry Eubanks, and Anne Hart for their assistance while I was in the field;

- Lord Ashby, Edwin Barber, Walter Kollmorgen, and Rebecca Roberts for their suggestions for improving initial drafts;

- Liz Barosko, Judith Kordahl, and Margaret Rasmussen for entering those drafts, plus a seemingly interminable string of revisions, in the word processor;

- Sally Myers, for her meticulous editorial advice; and

- Lori Brown, Hee-Bang Choe, Greg Chu, Carol Gersmehl, Sue Marien, Don Pirius, and Phil Schwartzberg for their cartographic skills.

Footnotes unfortunately seem to have been priced out of style, and I abominate endnotes and author-date citations, but it would be improper not to acknowledge my debt to certain published works that have been especially useful.

Essential reference works on the geography of the eastern United States include: Nevin M. Fenneman, *Physiography of Eastern United States* (New York: McGraw-Hill, 1938); William D. Thornbury, *Regional Geomorphology of the United States* (New York: Wiley, 1965); E. Lucy Braun, *Deciduous Forests of Eastern North America* (Philadelphia: Blakiston, 1950); Karl H. W. Klages, *Ecological Crop Geography* (New York: Macmillan, 1942); Ralph H. Brown, *Historical Geography of the United States* (New York: Harcourt, Brace and World, 1948); and William A. Koelsch, *Lectures on the Historical Geography of the United States as Given in 1933 by Harlan H. Barrows,* research paper 77 (Chicago: University of Chicago Department of Geography, 1962).

The two maps that have been my constant companions are C. P. Barnes and F. J. Marschner's *Natural Land-Use Areas of the United States,* 1:4,000,000 (Washington: U. S. Department of Agriculture, Bureau of Agricultural Economics, 1933) and Edwin H. Hammond's *Classes of Land-Surface Form in the Forty Eight States,* 1:5,000,000, map supplement no. 4, *Annals of the Association of American Geographers,* vol. 54 (1964). Any student of American agriculture relies heavily on the graphic summaries that have been published as parts of the various censuses of agriculture. They have been richly supplemented by Sam Bowers Hilliard's *Atlas of Antebellum Southern Agriculture* (Baton Rouge: Louisiana State University Press, 1984).

For the European roots of American agriculture I have relied most heavily on B. H. Slicher van Bath, *The Agrarian History of Western Europe, A.D. 500–1850,* trans. Olive Ordish (London: Edward Arnold, 1963) and M. E. Seebohm, *The Evolution of the English Farm,* 2d. rev. ed. (London: George Allen and Unwin, 1952). The classic work on the Pennsylvania Dutch is Walter M. Kollmorgen's *Culture of a Contemporary Rural Community: The Old Order Amish of Lancaster County, Pennsylvania,* Rural Life Studies, no. 4 (Washington: U. S. Department of Agriculture, Bureau of Agricultural Economics, 1942). It has been supplemented by James T. Lemon's *The Best Poor Man's Country: A Geographical Study of Early Southeastern Pennsylvania* (Baltimore: Johns Hopkins University Press, 1972); would that we had comparable dissertations on the historical agricultural geography of other important seedbed areas such as southwestern Virginia, upstate New York, and the Miami Valley of southwestern Ohio! I would be remiss not to mention Alfred L. Shoemaker, ed., *The Pennsylvania Barn* (Kutztown, Pa.: Pennsylvania Folklife Society, 1959) and Amos Long, Jr., *The Pennsylvania German Family Farm*, publication of the Pennsylvania German Society, vol. 4, (Breinigsville, Pa.: The Pennsylvania German Society, 1972).

Paul W. Gates's *The Farmer's Age: Agriculture, 1815–1860,* vol. 3 of *The Economic History of the United States* (New York: Holt, Rinehart and Winston, 1960), is superb on the evolution of agricultural institutions, although as a geographer I wish he had been a bit more interested in location. F. H. Marschner's *Land Use and Its Patterns in the United States,* agriculture handbook no. 153 (Washington: U. S. Department of Agriculture, 1959), is the classic work on land survey systems. Norman J. W. Thrower's *Original Survey and Land Subdivision: A Comparative Study of the Form and Effect of Contrasting Cadastral Surveys,* Association of American Geographers monograph 4 (Chicago: Rand McNally and Co., 1966), describes the impact of these systems on Ohio. William T. Utter's *The Frontier State, 1803–1825,* vol. 2 of *The History of the State of Ohio,* Carl Wittke, ed. (Columbus: Ohio State Archaeological and Historical Society, 1942), had good material on early Corn Belt agriculture, and Carl C. Taylor et al., in *Rural Life in the United States* (New York: Alfred A. Knopf, 1949), describe it at the end of World War II. Carl E. Leighty's "Crop Rotation," in *Soils and Men: Yearbook of Agriculture, 1938* (Washington: U. S. Department of Agriculture, 1938), pp. 406–430, is still the best discussion of crop rotations.

Mack N. Leath, Lynn H. Meyer, and Lowell D. Hill's *U. S. Corn Industry,* agricultural economic report no. 479 (Washington: U. S. Department of Agriculture, 1982) is detailed. The history of dairy farming is discussed in Joseph Schafer, *A History of Agriculture in Wisconsin,* Wisconsin Domesday Book, general studies, volume 1 (Madison: State Historical Society of Wisconsin, 1922) and William Francis Raney, *Wisconsin: A Story of Progress* (New York: Prentice-Hall, 1940). Ben A. Franklin, "From Womb to Tomb on the Chicken Farm," *New York Times,* May 27, 1979, is an excellent description of the broiler business.

The *Proceedings of the Tall Timbers Ecology and Management Conference,* no. 16 (Tallahassee, Fla.: Tall Timbers Research Station, 1982), discusses plantations in general. For particular crops, see Sam B. Hilliard, "Antebellum Rice Culture in South Carolina and Georgia," in J. R. Gibson, ed., *European Settlement and Development in North America: Essays on Geographical Change in Honour and Memory of Andrew Hill Clark* (Toronto: University of Toronto Press, 1978), pp. 91–115; John J. Winberry, "Indigo in South Carolina: A Historical Geography," *Southeastern Geographer,* vol. 19 (1979): 91–102; Charles F. Kovacik and Robert E. Mason, "Changes in the Sea Island Cotton Industry," *Southeastern Geographer,* vol. 25 (1985): 77–104; and T. J. Woofter, Jr., *Landlord and Tenant on the Cotton Plantation* (Washington: Works Progress Administration, 1936).

Commodity specialists in the Economic Research Service of the U. S. Department of Agriculture prepared a series of extremely useful background papers on specific programs in 1984 to facilitate congressional deliberations on the 1985 farm bill. These papers were published as Agriculture Information Bulletins 467 (wheat), 468 (tobacco), 470 (rice), 471 (corn), 472 (soybeans), 474 (dairy), 476 (cotton), and 478 (sugar). I also used Bobby A. Huey, *Rice Production in Arkansas,* circular 476 rev. (Fayetteville: University of Arkansas Division of Agriculture Cooperative Extension Service, 1977); R. D. Beaux et al., *Culture of Sugarcane for Sugar Production in the Mississippi Delta,* Agricultural Handbook no. 417 (Washington: U. S. Department of Agriculture, Agricultural Research Service, 1972); and *Citrus Industry of Florida* (Tallahassee: Department of Agriculture, 1957). Robert N. Ford, *A Resource Use Analysis and Evaluation of the Everglades Agricultural Area,* Research Paper no. 42 (Chicago: University of Chicago Department of Geography, 1956), is a useful study.

I have used county soil surveys whenever they have been available. The older surveys generally have been more useful than more recent ones, but of course the ideal would be to have both.

Index